Location, Location, Location
A Plant Location and Site Selection Guide

Location, Location, Location
A Plant Location and Site Selection Guide

Marcel De Meirleir, PhD

Taylor & Francis Group

NEW YORK AND LONDON

First published by
The Haworth Press, Inc.
10 Alice Street
Binghamton, N Y 13904-1580

This edition published 2011 by Routledge

Routledge
Taylor & Francis Group
711 Third Avenue
New York, NY 10017

Routledge
Taylor & Francis Group
2 Park Square, Milton Park
Abingdon, Oxon OX14 4RN

© 2008 by The Haworth Press, Taylor & Francis Group. All rights reserved. No part of this work may be reproduced or utilized in any form or by any means, electronic or mechanical, including photocopying, microfilm, and recording, or by any information storage and retrieval system, without permission in writing from the publisher.

PUBLISHER'S NOTE
The development, preparation, and publication of this work has been undertaken with great care. However, the Publisher, employees, editors, and agents of The Haworth Press are not responsible for any errors contained herein or for consequences that may ensue from use of materials or information contained in this work. The Haworth Press is committed to the dissemination of ideas and information according to the highest standards of intellectual freedom and the free exchange of ideas. Statements made and opinions expressed in this publication do not necessarily reflect the views of the Publisher, Directors, management, or staff of The Haworth Press, or an endorsement by them.

Cover design by Kerry E. Mack.

Library of Congress Cataloging-in-Publication Data

De Meirleir, Marcel.
   Location, location, location : a plant location and site selection guide / Marcel De Meirleir.
     p. cm.
   Includes index.
   ISBN: 978-0-7890-1966-0 (hard : alk. paper)
   ISBN: 978-0-7890-1967-7 (soft : alk. paper)
   1. Industrial location—Handbooks, manuals, etc. 2. Industrial sites—Handbooks, manuals, etc. 3. Business relocation—Handbooks, manuals, etc. 4. International business enterprises—Location—Handbooks, manuals, etc. 5. Industrial location consultants—Handbooks, manuals, etc. 6. De Meirleir, Marcel. I. Title.

HD58.D416 2007
658.2'1—dc22

2007015505

CONTENTS

Acknowledgments ix
   The Pre-Chicago Era ix
   The Academic Period at the University of Chicago x
   Returning to Belgium xi
   The Plant Location International Period xii
   The Academic Career (1975-1990) xii

Introduction 1

Chapter 1. General Principles of Industrial Location 7
   The Base Philosophy: The Purpose of an Industrial Location Study 7
   Industrial Location 11
   Techniques Used in Industrial Location Studies 12
   Changes in Methodology and Approach in Recent Years 21

Chapter 2. The Project Definition: Basic Requirements 33
   The Ten Main Intangibles 41
   The BERI Rating 42

Chapter 3. Critical Factors Analysis 43
   The Critical Factors 43
   Locational Feasibility Studies 53
   Mergers, Joint Ventures, and Acquisitions 53

Chapter 4. The Economic Analysis—Part I: Analysis of the Nonrecurring Cost Factors 55
   Calculating Investment Costs 58

Chapter 5. The Economic Analysis—Part II: Analysis of the Recurring Cost Factors 69
   Labor Costs 69
   Transportation Costs 70
   Utility Cost 72

Taxation	74
Tax Incentives	77

Chapter 6. Analysis of Intangible Factors — 81

The Investment Climate	81
The Social Climate	82
The Political Situation	84
The Economic Situation	87

Chapter 7. Other Elements — 91

Financing "The Investments"	91
Cash Flow Tables	93
Industrial Location and Ecology	94
In Conclusion	95

Chapter 8. Community and Site Analysis — 97

Geographic Aspects	97
Socioeconomic Aspects	98
Site Location Process	100
Community Analysis	101
Site Analysis	104
Infrastructure	106

Chapter 9. Decision Making — 109

The Weighing of Factors in Location Decision Making	109
Drawing Conclusions and Making Recommendations	111
The Use of Consultants	115
In Conclusion	118

Chapter 10. Locating Warehouses, Research and Development Centers, Headquarters, and Call Centers — 121

Warehouses	121
Research and Development Laboratories and Research Centers	122
Headquarters	124
Call Centers	125

Chapter 11. Attracting Investments — 129

    Regional and Economic Balance — 129
    What Industries to Attract: Matching Offer and Demand — 131
    How to Cooperate with Industry — 134
    The Ten Commandments for Regions Desiring
      to Attract Investments — 135

Chapter 12. Industrial and Development Studies — 137

    The European Union — 137
    Other Areas in Europe — 139
    The Euro — 139

Chapter 13. Industrial Location As a Science and Its Role in the Future — 141

    Industrial Location As a Science — 141
    Location Studies and the Future — 142

Chapter 14. The Consultant's Role in Site Selection — 145

    Site Selection Principles — 145
    Who Seeks Sites? — 146
    When a Company Undertakes Its Own Location Study — 147
    How Does the Consultant Come In? — 148
    The Six Major Qualities of a Location Consultant — 148
    Incentives — 150
    The Final Recommendations — 151
    Implementing the Decision — 151
    In Conclusion — 152

The Language of Location — 153

List of Stories — 155

List of Location Tools — 157

Index — 159

ABOUT THE AUTHOR

Marcel De Meirleir, PhD, is CEO of Business Location International. He is internationally recognized as a leading authority on business location services with direct experience supporting investment projects by many top United States companies. He is also the founder of Plant Location International, an international business consulting service. He has been involved in over 1,500 location studies worldwide and has served as an advisor to the boards of many leading corporations. Dr. De Meirleir has been a well-respected professor at various universities across the United States.

Acknowledgments

This book is the result of the confidence earned from managers of hundreds of American, European, and Japanese companies whom my colleagues and I advised in the location decision. They trusted the capabilities of my company with investments that sometimes involved billion-dollar sums. Those companies learned a lot by working with us; and we learned a lot from them.

This book is a product of a half-century of academic and professional life. Recalling all the business leaders who had an influence on my education, work, and experience is an impossible task. Nevertheless, my partners and friends advised me to give readers the benefit of my exposure to some of the most outstanding personalities of the academic and business world during the past five decades.

THE PRE-CHICAGO ERA

After four years at the famous Belgian University of Louvain and a master's degree earned in Economic Geography, I was eager to visit the Tennessee Valley Authority (TVA). I had read a lot about the TVA and seen movies about it. I was told, however, that the leading scientists who had been involved in developing the TVA were by that time teaching at the University of Chicago. Impressed that the University of Chicago could attract such outstanding talent, I applied for admission there and was accepted. How did my dream come true? A number of outstanding men came to my rescue and made it possible for me to attend that great university.

I focus on two of those men, both of whom had brilliant careers in the politics of my country, Belgium, and of Europe. Anton Spinoy was the mayor of the city of Mechelen. Later he became minister of defense, then minister of economy, and eventually vice prime minister of Belgium. He introduced me to Paul-Henri Spaak, the famous

European who, with De Gaspari of Italy, Schuman of France, and Adenauer of Germany, was at the cradle of the founding of the European Union. I owe a lot to the support of Mr. Spaak, who held positions as the first president of the United Nations, then secretary-general of NATO, as well as prime minister and longtime minister of foreign affairs for Belgium.

When Mr. Spaak learned about my desire to study at the University of Chicago and knew that I had no backing for a scholarship, he persuaded the Belgian American Education Foundation (BAEF) to give me a special scholarship that was named after him. He also gave me three letters of introduction: one for Professor Compton, president of the University of Chicago; one for the Belgian Ambassador to the United States in Washington, DC, Baron Silvercruys; and one for the consul general in Chicago. The first recommendation turned out to be a godsend because Dean Strozier subsequently decided to award me tuition-free scholarships for nearly three years. It was also Mr. Spaak who arranged for my meeting with Winston Churchill shortly after labor leader Clement Attlee defeated him. Mr. Spaak had asked me to take some documents to Checkers, the Churchill family estate, in London.

THE ACADEMIC PERIOD AT THE UNIVERSITY OF CHICAGO

As mentioned previously, the introduction letter from Mr. Spaak did wonders for helping my admission to the University of Chicago. That letter also opened the door for my introduction to Nobel Peace laureate Dr. Compton, Dean Strozier, and Chancellor Robert M. Hutchins, who signed my PhD degree three years later.

I was well received in the geography department by both professors and students, most of whom were former GIs and had fought battles in Europe during World War II. In addition to young Professor Chauncey Harris, two other men had a profound effect upon my academic education: Professor C. Colby, the chairman of the geography department, and Professor Robert Platt, my dissertation mentor.

While taking a full-time load of classes for my geography major, I was allowed—in an exception to the rule—to take two minors: economics, then taught by Professor Senator Paul Douglas and Professor Milton Friedman (another Nobel laureate), and planning. The latter department permitted me to take courses in physical, economic, and social planning given by Professors Meyer, Perlof, and Ludlow, all of whom were formerly involved in the TVA, and had been the impetus for my application to the University of Chicago.

My doctoral dissertation on the relocation of industry in Chicago was made possible by Professor Harold Mayer—the research director of the Chicago Plan Commission under Commissioner Carl Gardner who hired me as his assistant. That team helped me to interview some 120 executives in Chicago, drew all of my maps, illustrated the thesis, and gave me a farewell cartoon—something I cherish very much.

RETURNING TO BELGIUM

The mayor of Mechelen helped me make the trip to Chicago so I could study there. To return the favor, I assisted him in dealing with a cruel 18.4 percent unemployment problem. He encouraged me to start attracting foreign direct investment (FDI), especially from the United States. I succeeded in that quest, so in a few years unemployment fell from that very high level of 18.4 percent to 4 percent of the active population. Among the many still operational companies that started manufacturing operations during that period are Procter & Gamble and Dupont de Nemours.

When I returned to the United States ten years later, in 1960, I lived in New York City. That is where I learned to understand the value of offering advice to expanding companies and how to apply the lessons I had learned from my studies and from recent practical experience. However, if I were to set up an operation to help companies relocate their operations or establish new ones, how might I find customers for the service?

The answer came one day when David Rockefeller, chairman of Chemical Bank, invited me to the Business Club on Wall Street. When I asked him the question, "How can I find clients?" he took me to the window of the thirty-third floor and said, "Look down on the street. Do you see all those ants? They are people. Go down there, tell them your story, and I am sure that one of them will need your advice."

THE PLANT LOCATION INTERNATIONAL PERIOD

During a thirty-year period, when my company, Plant Location International, conducted more than 1,500 projects, I met quite a few very interesting managers. The experience of working with those managers had a significant impact on my thinking. The relocation of Covent Garden Market, one of the most famous places in London, was brought to me through an introduction by Sir Charles Forte, the hotel guru, to Minister Arthur Soames, Winston Churchill's son-in-law, and to the general manager of Covent Garden Market, Frank Hollins. Minister Soames defended me against the Lords, who protested that the job was being given to a foreigner.

My operational contacts, whom I spent time with during field trips to potential sites, were unforgettable and rich in experience and wisdom. Mr. Jack Welch was the European manager of the Plastics Division when the decision to locate a factory at Bergen op Zoom in the Netherlands was made. I worked for Mr. Henry Ford II on many projects, first in Belgium, then later in the United Kingdom and France.

Unforgettable also were my discussions in Paris with President Georges Pompidou and Prime Minister Chaban-Delmas, as well as with the mayor of Bordeaux (where the transmission plant for Ford was built); Bruno Kreisky—later president of Austria—Mário Soares the president of Portugal when the Oporto studies were underway, and President Miguel De la Madrid of Mexico, when I was looking for sites for American pharmaceutical companies and studying the impact of the Mexico-United States border industrialization.

My company conducted more than twenty studies for General Motors (GM) Corporation. GM executives whose friendship and advice were invaluable included Alex Cunningham, general manager of Opel in Ruesselsheim, Germany, Gene Ptak, manager of manufacturing in Detroit, and Tom Parent, manager of Argonaut, the real estate division of General Motors. Mr. Chaplin, CEO of Woolworth United Kingdom, was impressed with our data gathering.

THE ACADEMIC CAREER (1975-1990)

While spending fifteen years as a professor at the Free University of Brussels (VUB), I was permitted to set up a postgraduate program in industrial location and regional development. During this period, about 500 students from 28 countries participated in the program. My gratitude especially goes to Professor Piet Frantzen, the dean of the faculty in economic, social, and political sciences (ESP) who gave me the honor of participating in the 1992 mission to China along with six university colleagues. The resulting publication from that mission drew attention to the booming economic development in China, which is still witnessed. I would like to make special mention of the support I received from Karel Van Miert, another great European, who did a magnificent job as a two-term member of the European Commission. The value of his engagement to spread the gospel of location cannot be measured.

With this background, the reader now will understand why sound location advice is the result of *experience.* For a long time, I hesitated to write this book knowing that the world of economics and politics moves very fast. However, I am sure that the principles I drew from location studies and regional development remain unchanged today, and I am confident those same principles will remain unchanged in the future.

Special thanks go to my family, my wife especially, who has seen me work many nights on this manuscript. My family will be amazed to see this book, promised ten years ago, come to completion.

Finally, I am grateful to my friend, Professor Myron Miller, who has acted as a co-author. He spent considerable time cleaning up the text, gave me valuable advice to make the book easy to understand, and helped select my best stories to illustrate various ideas described in the book.

Introduction

Yes, dear reader, you guessed right: this book is about the location of facilities. Repeating the word "location" three times is meant to stress the importance of the locational decision in management processes. This book is based on a half-century of experience by myself, and probably should have been written many years ago. Nevertheless, maturity is a great attribute, and experience is essential for success in making critical decisions. Indeed, knowledge of location decisions can only be acquired by meeting three criteria: *experience, experience,* and *experience!*

I and my associates have dealt with more than 1,500 location cases worldwide, so I felt that the time was right for me to preach the gospel of "location." The purpose of this book is not to be a bible, rather, it provides an analysis of the evolution of locational problems from the end of World War II until the present time. This book also provides the reader with an ability to face the changes in the approach to location problems as we enter the twenty-first century.

In spite of the importance of location decisions for management, few books have been published on the subject, although many articles have been written about it. My associates and I have attempted to introduce the subject of location to universities for many years and have finally achieved some success in the past fifteen years. By and large, the main courses in the field of location have been reduced to secondary unimportant topics in MBA programs. Because of the scope of knowledge required to manage location decisions, the topic of location often does not fit into the traditional management disciplines. Nevertheless, senior managers still look for courses where they can learn how to avoid the many costly mistakes that occur daily in location decisions.

Who are the people involved in location decisions? This is a difficult question to answer because the situation varies from company to company. Location decisions may be managed by the board of directors, a vice president, the manufacturing buyer, or any member of the advisory staff.

The location decision is one of the most difficult for management to make, yet it has very far-reaching consequences. The way a location decision is made can mean the difference between success and failure.

Quite often, one can read about success stories that indicate that—all factors being equal—the result was due to three things: location, location, and location.

The time from when a concept is developed for a new manufacturing facility until the time when actual production begins at the new plant is usually three to five years. If at the end of this cycle it is apparent that the decision was wrong, substantial losses are being incurred, and production is uneconomical, who is to be blamed? Who will be replaced because of the faulty decision?

When irregularities occur in the company's accounting, or if the annual report doesn't look good, the finance manager is fired. When a company loses cases in court, the legal counsel's head is on the block. When the factory produces a poor quality product, the manufacturing manager is replaced. When a new product doesn't sell, the sales manager and/or the publicity manager must look for another job. However, who gets the blame if, after starting operations at a new site, it turns out to have been the wrong site? The president of the company or the entire board of directors gets the blame; but neither will be fired.

Ignoring warnings about potential mistakes in choosing a location is a luxury that management can ill afford. If the warnings are ignored, mistakes will be made in the future as they have been made in the past. After providing guidance in more than 1,500 cases, we have yet to find a situation where we have failed, so we are sure we are on the right track with our advice.

The purpose of this book is to pass on some of this experience to any company that is making a location decision. Perhaps we should apologize in advance that some of the more theoretical cases might be a bit less interesting than the real cases that will be used. However, the theoretical cases all carry a very useful message and valuable principles. Since no one likes to be considered a fool, the citations will be masked. Only those who were involved, though in most cases they are no longer around, will recognize the stories.

Location is a two-way street. It deals with offers and demands. Two parties are always involved: those who "offer" are the authorities (local, national, regional, and international), while those who "demand" are the investors. This negotiation leads to a compromise in the widest sense of the word. Both sides of the negotiation will benefit from the knowledge of what the other side offers, should offer, or would like to receive.

Plant Location in Spain[1]

This story happened in the mid-1960s. A company president was approaching retirement age. To celebrate his fortieth wedding anniversary, the president and his wife decided to make a one-month trip around the world. He had foremost in his mind the need for a manufacturing plant in Europe.

In early May, the couple landed in Madrid, Spain. The weather was excellent. It was the celebration of San Isidro (the patron saint of Madrid), so very little work was done that week. The bullfights were colorful and the crowds cheerful. The president's wife fell in love with Madrid. The company's representative in Madrid knew all the best restaurants and other attributes of the city. He took the celebrating couple to see the flamenco dancers and the bullfights, and to enjoy the famous seafood restaurants. During the conversation, the idea of a new factory in Europe was raised. The president's wife suggested to her husband, "Why not in Madrid?"

As the president later told me, from the moment that his wife asked that question, everything moved very fast. The local representative in Spain had a brother in the real estate business, and that person found a site for the factory. Another brother produced the drawings and the building permit in a short time. The representative had a cousin, a notary public, who processed the deed of the land. He also had a friend who was a top business lawyer in Madrid and that person provided legal advice. Finally, his best friend ran an accounting firm that would be able and willing to shield the company from paying too much in taxes.

We were surprised to hear of this story, as was the vice president of the company, who had responsibility for site location. We had discussed with him at great length the basic requirements of the project, particularly the operating costs of the new plant.

An important factor in this location decision was transportation costs. Indeed, for this project, all raw materials were to be shipped to the port of Barcelona, then trucked to Madrid—a distance of 600 kilometers—on poor, narrow, and winding roads. At that time there was no express highway between the two cities. Further, the Madrid area provided a market of less than 5 percent of the output of the factory, while the other 95 percent had to be shipped back to Barcelona and hence to the rest of the European market.

The factory was built and operational, but sustained great losses over a period of a few years. We discovered while checking on the progress of the project that the factory had been sold to another of our clients at a very low cost. The president who had made that decision retired shortly thereafter. No one took the blame for the decision to choose the wrong location. Location can make or break any operation if the cost factors are misjudged or if the intangibles turn out to be negative.[2]

[1] This plant location story in Spain is just one of the many incidences involving wrong locational decisions. No names have been exposed. These stories are generally swept under the carpet; however, we want our readers to learn from these situations.

[2] The influence of "woman," in this case, the president's wife, was publicized in an article titled "Don't let your wife select the plant site." Though a wife had a great influence on the man in the Spain scenario, it is now the case that a man or a husband could strongly influence a wife who is the president of a company.

This book, therefore, will not split the problem of the "marriage" between man and wife (region and investor) into separate approaches; rather, it will deal with the connection between the two as one unit. A happy marriage will not lead to divorce if the criteria or the requirements of both parties are met. Although we do not have a crystal ball, nor do we practice clairvoyance, we will nevertheless attempt to foresee the main factors that will affect "location, location, and location" in the future.

This brings us, before starting the systematic analysis of locational decision making, to the following question: Who is involved in locational decision making?

In preparing to make the locational decision, the board of directors, represented by the chief executive officer, is ultimately responsible for making the final decision for an important investment on a given site. The preparations for the decision to be made by the board of directors are conducted by staff members who are the facility planners or by a team appointed to make the assessment. There is no fixed rule as to who makes the decision; however, since the 1970s, companies have insisted on more precise information and sophisticated techniques to avoid making the wrong decisions.

The facility planner as a function is a new concept, started about forty years ago. That person can be a quantity surveyor, an architect, a civil engineer, and a real estate professional. The responsibilities of a facility planner are usually a combination of several functions since the frequency of and the need for locational decisions are rather small or nonexistent in years of economic recession.

Large, expanding companies have established a special staff that is at times composed of a combination of professions, including the management of acquisitions and what is commonly referred to as "industrial real estate" specialists.

Emphasis has shifted to concepts for planning for these decisions, seen in such titles as "real estate planning" and/or "long-range planning." In recent years expansion planning has been fine-tuned to give greater weight to the location factor. Engineers tend to investigate more intensively the physical site aspects of the location, and less, the economic aspects. Facility planners trained in real estate tend to concentrate on the value of the property, the layout, and other important property-related factors.

Training in location was provided from 1975 until 1990 at the Flemish University of Brussels (VUB), economic, social, and political faculty (ESP) in a two-year postgraduate program. In view of the absence today of such training, many methods are used in gathering and presenting the data needed to make a valid and wise location decision.

Great merit must be given to The International Development Research Council in Atlanta (IDRC)[1] and its founder, Mr. McKinley Conway, as well as the fine group of facility planners of the major American corporations. The IRDC, through its publications and biannual three-day conferences in the United States, provides a platform for facility planners and prospective communities to meet. The supply and demand sides have an opportunity to get to know more about each other and thus come closer to realizing the "ideal marriage" between the investor company and the community where new sites will be located.

Preparing for a locational decision is very complex. There is a need for experience in dealing with a large variety of projects, company

---

1. IRDC has since merged with Nacore into CoreNet Global.

management, company policies, ever-changing investment climates, quickly evolving economic realities, changing personalities in countries and regions, and a need to identify a large number of the intangibles that affect investment projects.

We wrote this book to pass on the experience gained through our work and to honor the hundreds of company managers who showed their confidence in us by hiring our services, the facility planners of major corporations we've worked with, and the regional and local personalities—responsible for attracting industry in their areas—with whom we have collaborated over the past four decades. We also honor the nearly 500 students from more then twenty-eight different countries, who were eager to learn from our experience and were keen to apply that knowledge for application in their own countries and companies.

Chapter 1

General Principles of Industrial Location

THE BASE PHILOSOPHY: THE PURPOSE OF AN INDUSTRIAL LOCATION STUDY

Why do companies invest in production facilities, warehouses, laboratories, research units, or in a headquarters operation? The answer: to satisfy a need, to make money, to serve a market, to capitalize on a patent, or to distribute a product.

The purpose of a location study is to determine the area and the site at which the projected operation and investment can be carried out under optimal conditions, with the best monetary return, and with the least number of problems.

To arrive at the best possible location recommendation, a number of steps must be followed. These steps are enumerated hereafter, with the objective that each chapter contains a detailed description and analysis of one step.

In Chapter 2, a beginning point is established for addressing the *basic requirements* of a location project, using an all-purpose location questionnaire to complete the *project definition*. From this questionnaire, a *critical factor* must be determined. In Chapter 3, a critical factor is defined as a locational factor which makes it uneconomical to locate at a specific site or in a specific location. A critical factor is a determining location factor that dominates all other factors and plays a prohibitive role in site selection.

For example, a factory requiring large amounts of water should not be located in desert areas where water is scarce. The need for a canal to transport with barges will exclude many areas from the locational pattern of the site search. If labor cost carries the highest operational cost in a labor-intensive project, it will be the *determining* factor in the location decision, although all other factors must also be analyzed and studied. The listing of basic requirements will help determine whether a project is capital or labor intensive.

In Chapter 3, the *area of search* is defined as the area in which the communities and sites will be assessed to satisfy basic requirements in an optimal manner. Most areas of search will be eliminated on the basis of the initial factor analysis, the determining factor or the requirements initially determined by the company, the nature of the investment (labor-intensive or capital-intensive), and other considerations such as proximity to markets, sources of raw materials, the need to obtain maximum investment incentives, and so on.

At this point in Chapter 3, we are ready to establish the *physical fit* and matching exercise, which will play such a dominant role in the *community approach to industrial location.* When the locational fit is established, the analysis of the preferred regions, communities, and sites can start. All national-, regional-, and site-related factors have been dealt with so that conclusions can be drawn for this part of the analysis.

Chapter 4 and Chapter 5 deal with the *economics* of the projected investment. We shall look at the *investment cost,* a *nonrecurring location factor* that includes the one-time investment in land (site), buildings, and equipment, from which will be deducted the cash grant obtained from the governmental authorities. We shall also investigate the *operational cost,* a *recurring location factor* that includes labor cost, transportation cost, and the cost of utilities (including environmental protection) and taxes. The value of the locational cash flow table will be demonstrated in these chapters.

In Chapter 6, the major intangible location factors are analyzed and their respective importances are demonstrated. After all, these factors make up the *investment climate* in which the company is going to operate in the country selected.

In Chapter 7, the role of specific elements, notably ecology permits and costs and raw materials and services, in the locational decision is assessed. Feasibility studies are to be understood and evaluated before final conclusions are drawn.

In Chapter 8, options are provided for locations now available to management. The communities and sites are fully analyzed and evaluated in light of the project requirements. Chapter 9 deals with decision making. In addition, recommendations are provided for the implementation of the location study. Let us be clear at the outset. The "ideal" location does not exist. Neither does the "ideal" person exist. All preferred sites have their merits and their drawbacks. Any location decision is the result of a compromise of factors.

In forty years of handling more than 1,500 studies, only once did we come to the conclusion that our number one candidate was "it." We could not find one single drawback. Unfortunately, the project, which involved an electric motor plant for the automotive industry, was never carried out due to the energy cost crisis of the early 1970s.

The location decision for a production facility is one of the most important decisions that management faces. To make a mistake and decide on the wrong location, a location where the policies of the company cannot be carried out or where operating losses are detected only when it is too late, can have serious consequences. It is very difficult to identify the individual on the team who was instrumental in making that wrong decision.

Only since the end of World War II and the invasion of Europe by American companies in the mid-1950s and the "golden 1960s" did companies choosing locations for their expanding facilities start to give plant location adequate attention.

Initially, when the demand for new facilities was large, competition for business was slim and opportunities were plentiful in the countries devastated by World War II in Europe. The choice of a country and a region for new facilities was determined by fear of the unknown, the lack of knowledge of the European languages, and confusion about the differences in regional customs and social attitudes. The choice of location was also affected by the growing influence of the labor unions, the effects of the Cold War, and the spread of the politically left wing ideas.

These attitudes, however, have changed gradually as more and more companies build plant facilities in Europe. Examples of success have been the key to expanding the number of companies establishing facilities in Europe. Once a large American multinational moved into Europe, others followed its example. Mechelen, a city in Belgium equidistant (sixteen miles) from Brussels, the capital of Belgium, and Antwerp, the second largest port city in Europe, used this phenomenon to the fullest. I had moved to Mechelen after leaving Chicago, anticipating that location as a focal point for American companies desiring to expand their operations into Europe. On my advice, the mayor of Mechelen encouraged contact with American companies.

Land for development was acquired by the city of Mechelen and the infrastructure needs of industry were also in place. That land was located close to the workers' homes. Car owners were still not numerous at that time, and the bicycle was the primary means of transportation. The land was sold at cost to the investing companies and, for the first time, the city provided incentives in the form of infrastructure: roads, water, sewer systems, and streetlights.

Procter & Gamble built its main European plant in that area, followed there by the paint plant of Dupont de Nemours, Inc. Success breeds success and companies such as those that produced chainsaws (McCullough), electric connectors, and so on soon followed. Mechelen also became an important car assembly city: Peugeot, Mercedes, Austin, Standard Triumph, and others built small car assembly plants.

We cautioned the prospective companies that these plants had only a limited number of years to function profitably. High duties on imported cars and other trade restrictions would provide a barrier to competition for some time, but not forever. These plants benefited Mechelen and Belgium by creating well-paying jobs. Unemployment, which had been at 18 percent after the end of World War II, was reduced to less than 4 percent within the first decade after the war. Technical schools were opened in cooperation with the new plants, and workers were trained for the higher-level skills needed in those factories. These new, higher-paid jobs raised wages and standard of living for those workers.

Unfortunately, the negative effect of the new large industrial companies was the gradual disappearance of two of the area's traditional industries: hand-made, sculptured home furniture and the canned food industry for which the city and the region had become famous. When the Common Market, which had resulted from the Treaty of Rome, came into being, the small home car market no longer justified a closed market, so the factories that produced those cars in Mechelen were closed, one after the other. In the meantime, however, broader industrial development was well underway.

The strategic location of the city of Mechelen, our basic philosophy of industrial location, the creation of industrial parks, cooperation between industry and public authorities as early as the 1950s, and the change in attitude between workers and unions allied to a more harmonious labor climate and made the city a pioneer in industrial development. Our knowledge of American industry's needs and requirements, our prospective philosophy and language ability, and our three years at the University of Chicago helped the city of Mechelen develop a model for industrial development which was soon copied by other cities and regions in Europe.

The industrial location model, established at Mechelen, created an awareness that investments are made to serve an existing and potential market and allow the investor to make money while creating jobs and providing purchasing power as well as a better standard of living. The development of this type of model is consistent with the basic philosophy of sound location policies.

By the 1950s, Dupont de Nemours required a justification of the location decision by means of a study, in which we applied for the first time the basic principles upon which our later work was based. In the early 1960s, the assembly plant location studies for the Ford Motor Company led to the building of the Genk Taunus plant in Belgium. The plant, built in 1961, is still employing more than 12,000 workers directly and more than double that amount indirectly. That project demonstrated the value of thorough evaluation of the locational factors described in this book.

INDUSTRIAL LOCATION

Two phenomena of industrial location concentration became apparent in continental Europe in the post–World War II era.

- Area concentration
- Supplier concentration

Area Concentration

During the reconstruction of the war-torn regions of continental Europe that began in the early 1950s, awareness of and the need for land planning and city development led to the introduction of general policies. Industrial districts, industrial parks, or whatever name was given were already established and known in the United States and the United Kingdom. In continental Europe, industries were located in or near the center of cities and towns, especially those in which female labor was predominant. Worker mobility was low and public transportation underdeveloped. Private cars were scarce and expensive to own and maintain.

The bombed cities of West Germany were not only rebuilt but also rezoned and restructured. Working hours were long (eight to ten hours a day) and leisure days were few. The work-to-home relationship was close particularly because workers, both male and female, began employment at an early age. Growing up in the 1920s and 1930s, we saw how these socioeconomic features dominated relationships among family members.

In the second half of the twentieth century, considerable land-use restructuring took place, which one might now consider self-evident. Are we satisfied today with the results of that restructuring? Can we learn from our mistakes? What kind of evolution might we experience in the twenty-first century?

Supplier Concentration

Dispersion of industry over the territory gradually dominated reconstruction of the European regions. Since we recommended the location of the Ford Motor Company's Taunus plant in Genk, Belgium (located in Limburg Province) in 1961, several other automotive suppliers began looking for sites in that province. The first of these was the Monroe Auto Equipment Company (now a part of the Tenneco Group), a producer of shock absorbers, then owned by the McIntyre family (the McIntyres were the owners of Monroe, which they sold to Tenneco). By the end of the twentieth century, the trend toward the clustering of companies had gradually evolved.

TECHNIQUES USED IN INDUSTRIAL LOCATION STUDIES

When we established Plant Location International in the early 1960s, the most difficult challenge was to win the confidence of management in American companies that wanted to expand into Europe. Most companies had the idea that one of their engineers, usually ready to retire, would be the best person to search for a site in Europe. There seemed to be no objection to the fact that the man had little experience in dealing with the social and economic mosaic of the European regions (not to mention the countries). It was assumed that the most successful representative or dealer for the company in Europe would be very helpful to the American representative. Of course, when the companies chose England for their entry into Europe, language was no obstacle.

Regional development organizations and the many city offices created to attract direct foreign investment only came into being by the middle of the 1960s. Their role and influence will be discussed later in this book.

Initial primitive techniques used in determining the best site for a project can be called, at best, amateurish. As a result of using primitive techniques, many expensive mistakes were made. Most of those mistakes were swept under the carpet when those responsible for the failures were quickly removed from their positions or were "kicked upstairs." The most shameful fact is that even today after forty years of practicing and preaching the gospel of "sound location studies," the evil of using amateurish techniques has not yet been removed.

Let us look at how location studies should be conducted. Before starting a location study, three very important aspects of a company's expansion plan should be examined.

- What are the company's short- and long-term goals and its policies to achieve them?
- What are the main requirements of the investment project, so one can know how the company makes money?
- What techniques are used during the investigation?

Company Policies

All companies want to grow, to increase their share of the market, and to make as much money as possible as a means to finance expansion and to reward its shareholders. How they make money, however, will determine how locational decision making is performed. The techniques and principles applied in location studies are valid for all investment projects.

However, 60 percent of all the criteria are specific to industry groups: the automotive industry, the chemical and pharmaceutical industry, and the electronics industry, for example. Thirty percent of the criteria have a geographical basis or are influenced by the timing of the project. Ten percent of the criteria are in most cases decisive criteria and relate either to company policy or personality factors.

In locating car assembly plants for General Motors or the Ford Motor Company, in which each project involved an investment of more than one billion dollars, the methodology used—and described in detail in this book—was the same. The differences in style and outcome were primarily affected by timing and preferred geographical areas. However, at times one strong-minded person could force the decision of where to locate a site in one direction or another. That was the case with some of Ford's expansion decisions in the 1960s. Although the influence of that one person did not lead to a scientifically justified decision, it had the benefit of speeding up the decision process and turned out to be quite financially rewarding.

The consultant and the team leader of the location study assignment must do their homework, initially by understanding the processes and decision policies of the client company's management.

The Main Requirements

The main purpose of the project requirements questionnaire is to evaluate and calculate *all* the factors affecting the locational decision. A model of such a "projects requirements" questionnaire is shown at the end of this chapter. This questionnaire has been revised almost fifteen times over the past forty years. Even at that, it is certainly not a finished product. Time and geography will make further adaptations necessary.

The headings, nevertheless, will remain, and include:

I. The Objectives

1. What kind of facility must be located (or relocated)?
2. What geographic area is the company considering? The study should point out that factors other than those suggested by the client *should* or *could* be taken into consideration if valid reasons become obvious in the course of the investigation.
3. The new facility should fit in with the strategic business development plans of the company. What are the client's views?

4. Various suggestions should be considered such as what the company thinks about locating close to existing competitors and the need for financial incentives. These could be *critical factors*. Many companies are somewhat reluctant to apply for incentives for a number of reasons, the two main ones being:
    - It is a fact that the reason for which incentives are given in a region is to solve economic, social, or ethnic problems of the region, all of which may affect the operations of the investor. *Remember that the investor wants to locate in an area where there are the least possible problems. The more problems and drawbacks, the higher the incentives offered.*
    - When applying for incentives, not only is the bureaucracy enormous (and expensive) but also the questions asked are of such a nature that answers given to the authorities could make public certain aspects of the company's technology and its business, which the company does not wish to expose, especially to its competition. Administration of public bodies is certainly not security-proof. Over the years we have witnessed many cases, in nearly all European countries, where information, which came from data requested by the governments and which was obtained when the company applied for incentives, has been divulged to competing companies. For the real value of incentives see "Incentives" on page 150.
5. Some companies want to locate facilities at a distance from competing companies to avoid pirating of labor and labor cost pressures, while some will want the proximity of a successful competitor. As an example of this first attitude (labor pirating): high-tech operations (such as a chip plant) requested specifically that their facilities be located outside the labor drawing area of its main competitors. An example of the second attitude (labor cost pressures) was the automotive assembly plant location, where the company preferred the proximity of a competitor operating two and a half shifts. The shift workers of the competitors tended to jump more easily to the client company if they were not satisfied. This strategy worked for our client company!
6. Companies tend to invest less and less in bricks and mortar. Instead, they like alternative investment formulas, such as leasing, which allow the company to keep its money for research and de-

velopment or equipment. In some cases, the companies want to benefit by investment approaches that gain from favorable tax laws.

II. Site and Buildings

1. *Area and shape.* While land cost was very low in the 1950s and 1960s, clients were advised to buy sufficient acreage to allow for expansion. Today a shortage exists of available sites in nearly all regions of Europe due to the expansion of industries in the last quarter of the twentieth century, the relocation of companies from the inner cities to the outskirts, and restrictions imposed by zoning. The intention of these restrictions is to create order in land use and to conserve land for agriculture, recreation, and natural resources.

   As a rule in past years, investment in land was about five percent of the total investment. This percentage has increased substantially in recent years and now reaches 25 percent for prime sites used by high-tech companies. The infrastructure cost of industrial sites should be investigated thoroughly, since it could reach large proportions if the authorities do not give subsidies or are unwilling to make an investment in the necessary infrastructure. The chemical industry is especially sensitive to infrastructure cost on greenfield sites (sites where there has been no prior investment).

2. *Special requirements.* An example might be the need for a prestigious location close to a well-traveled road for publicity reasons (such as the producers of consumer goods or providers of services who want to create awareness), or the proximity of access to a major highway or to an international airport.

3. *Transport facilities.* Water transportation in Europe has declined in recent years as more and better highways are built. The same applies to railroads. The nationalized railroad companies of the European countries have lost more and more traffic. The rail sidings of the 1950s and 1960s were praised as the "cheap" or "low-cost" solution, especially for bulk transport. Containerization and the development of the packaging industry, as well as the need for ever-increasing speed of delivery increased the

use of trucking to such an extent that highways converging on the major cities of Europe are gradually clogging up.

The need for parking space for company personnel has increased significantly. People avoid the use of public transportation services (the capacity of which has been declining) and make more use of private cars than ever before. Labor mobility is at a peak and must be investigated by the companies before siting facilities. Companies must take into account the improvement of public transportation facilities where they have been made.

III. Financial Assumptions

In order to calculate investment and operational costs, the purpose of which is to make a comparative analysis of the cash flow tables, a number of financial assumptions are needed. These assumptions will be projected to determine the *degree of importance* for each of the proposed sites under investigation.

1. *The projected initial investment* for the acquisition of the site, the construction (or purchase) of the buildings, the costs of equipment and tooling must be known. In many cases in which the client does not want to disclose the cost of equipment and tooling separately, an overall figure useful for the cash flow calculations and the total investment cost will then be given.
2. *The cost of raw material* is *never* needed for a location study since the siting of the plant will not affect the basic raw material cost but only the *transport cost* of the raw material. In many cases *basic point* or *delivery cost* will be offered by the suppliers. However, it is important to have a general idea of the products needed in order to consider *transport* or *handling* problems.
3. *Working capital requirements* such as cash to finance the labor bills, accounts receivable, and works in progress are useful parameters for preparing cash flows, especially when subsidies or other incentives (such as cheap bank loans) are offered.
4. *Annual sales projections* for the first five years of operations and projected operating income before taxes will help in the evaluation of the financial advantages of the sites.

While the investigation must identify available sources of financing in the proposed area of search, along with the initial investment and operating costs, a company may also consider arrangements with banks in the United States or elsewhere before embarking on an expansion into Europe.

IV. Products, Workers, and Distribution

1. Identifying products to be manufactured in the new plant and the geographic market to be served will help to identify potential areas of search.
2. Determining volume and weights of products and the mode of transportation will provide an understanding of their impact in the area in which the site search is directed.
3. Assessing the proposed distribution pattern will help in a calculated weighted center point of distribution, assist in determining the logistics of distribution, and keep costs as low and competitive as possible.

V. Raw Materials and Industrial Services

Evaluation of raw materials and industrial services requirements assist the investigation by establishing the relationship between the various sources of supply and transport. Availability of adequate industrial services in areas under investigation could reduce investments in maintenance, repair, and tooling shops.

VI. Labor or Human Resources

Knowledge of the company's requirements for human resources is increasingly becoming a necessity. The kind of skills, the amount projected, attitude toward shift-work, mobility of workers, and management union relations are all of vital importance in determining the likelihood of success of a new plant.

VII. Utilities

Reliability and cost of utilities are two factors affecting the location decision. Particularly after World War II, reliability became a problem in the newly industrializing countries. This situation, however, has improved gradually. Cost, though, remains an important element in evaluation. Denationalization and privatization of European utilities resulting from the unification of Europe are increasing competition, so utility costs are declining. Power, gas, water, and waste all must be studied intensively, especially for projects (such as chemicals and pharmaceuticals) in which large quantities of those utilities are consumed.

VIII. Effect on Ecology

In view of increased awareness of the preservation of life and nature, the questionnaire should be completed by assessing the impact of production on ecology: water, air, disposables, and waste. The cost resulting from numerous environmental protection laws weighs heavily on the location decision.

Unlike other location factors, ecology is at the same time *an investment cost, an operational cost, and an intangible factor.*

Tools

What tools are at the disposal of the location investigator in order to give the proper advice in selecting the "optimal" location? Notice that we refer to the "optimal" location and not the "ideal" or "perfect" location. Indeed, they don't exist.

A location is optimal given certain requirements and conditions in time and space that must be reevaluated periodically. A location is only temporarily sited. Products are developed, improved, and then they disappear. Methods of production change. Innovation constantly remakes society. The consumer's appetite is insatiable.

I. The Questionnaire and the Visit to a Similar Plant

Preparation of a questionnaire and a visit to a plant that is similar to the one under consideration are both done at the start of the investigation. We visited many hundreds of plants all over the world. It was the best experience one can have to qualify for making good location judgments. Too many consultants dealing with siting have never visited the types of plants for which they must determine the location.

II. The Interviews

Never go to an interview unprepared. Know with whom you will meet.

1. At the client's company:
   - If a *facilities planner* or *staff* is appointed, watch for a biased emphasis toward the real estate aspects of the project, while giving less emphasis to the operational economics.
   - If the *finance manager* or the *head of accounting* is involved, prepare the numbers (and their interpretation) very carefully. Also, don't lose sight of the intangible factors.
   - If the *chief engineer* (production and/or planning) is in your party, make sure you understand the company's management policies.
   - If you talk to the *president,* find his or her areas of greatest concern; otherwise your report will not be well received.
   - If you deal with the *sales manager,* stress logistics and market penetration but don't overlook locational costs.
   - If the *human resources* manager serves as liaison for your study, listen carefully to his or her needs, but brief him or her on the other aspects of the locational investigation.
2. In the regions:
   - In talking to the *regional development manager* or the manager's staff, remember that he or she *must sell that area to you and may distort or hide the truth.*
   - Interviewing the *mayor of the town* is the most difficult job. It is impossible for the individual to remain objective since he or she will insist that it is not necessary for you to look at other sites, since he has the best site available.
   - Questioning the *labor union officials* of the area will provide another perspective of the story, for they may say that an area has the *lowest labor costs* and the *best labor-management relations* (be attentive to the record of strikes in that area, and the possibility that there may be strikes).
   - The *management association secretary-general* will draw your attention to the problems one faces when operating in the area: labor pirating, the high cost of labor, absenteeism, and the negative attitude of labor unions.
   - *Utility companies* will offer the lowest *unit cost* but not indicate the total amount of a bill.

- The *employers* in the area that are your clients are the best sources of information.
- Somewhere among all those stories one must draw a realistic picture of the environment in which the new venture will have to operate.

III. Statistics

Collecting and using statistics is a way of being deceptive, if one chooses to do that. It is possible to hide the truth, distort the truth, or *use statistics to make a point.* Unfortunately, most statistics are either outdated, not comparable, or interpreted in favor of or against the proposed choice of location. Remember the saying "the glass is either half empty or half full." That wisdom can be applied to how statistics might affect a locational decision.

Avoid the use of statistics like the plague and make use of them only if you can verify the source as well as the method used in collecting the data. It is better to use *averages* and *orders of magnitude* in making locational recommendations. There can be no absolute truth possible in the interpretation of all the facts and figures gathered in the course of the investigation.

IV. Geography

Maps and charts are valuable tools for the locational factor analysis. One does not have to be a geographer nor a cartographer to use the materials available, but one must be able to make the correct interpretation and use illustrations of the findings provided in his or her report.

CHANGES IN METHODOLOGY AND APPROACH IN RECENT YEARS

Since Plant Location International (PLI) was sold to Price Waterhouse in 1989, major changes have occurred in the methodology and vehicles used by consultants in location studies.

Methodology

Personnel

Efficiency, training, and hiring and firing practices have now become *human resources* material.

Freight

Inbound, outbound, and interplant freight have been bundled into the new concept called *logistics*. The term "logistics" also includes warehousing and handling.

Location Factors

Tangible or intangible, calculable or not, location factors have been included in the evaluation as *benchmarking*.

Consultants

As a consulting company, we previously covered nearly three-fourths of the investments (American and Japanese) into Western Europe between 1960 and 1989. Four years later, in 1993, interviews with important customers revealed that large corporations in particular had returned to in-house studies and evaluations by their engineering, facilities (real estate) group, or the finance department.

Although Plant Location International had become a household name for all aspects of location, each of the larger consulting companies diversified from accountancy and tax work to general management advice, or to some kind of real estate and/or location advice under different headings, as follows.

- International Location Advisory Services (ILAS) of Ernst & Young
- Business Location Services (BLS) of Arthur Andersen, which disintegrated with the fall of Arthur Andersen
- Global Location and Expansion Services (GLES) of KPMG
- International Investment Group of Price Waterhouse, since sold to IBM

- Plant Location International (PLI) of PWC, since sold to IBM and later bought back in March 2005
- Rene Buck and his company, Buck Consultants International, which represents a continuation of our basic approach and which he has applied successfully from the beginning of his company's activity. His studies are in line with the practices of Plant Location International

Confusion

One of the reasons why large companies are reluctant to use consultants that "specialize in location" is the confusion created by some of those companies regarding whether the consultant is serving the interests of the demand (the investor) or the offer (the authorities).

Regions have often tempted consultants by offering large "sweeteners," that is, study assignments provided if success (which means investments) is guaranteed. This is a one-way street. Indeed, the consulting firm that submits to such practices is not only behaving in an unethical way, but also losing the confidence of investors, the people to whom that practice should have its first loyalty. The consulting firm that sells framed (similar) studies to several regions (in Europe and elsewhere) is doomed to lose its credibility.

It is entirely possible to make an honest evaluation of a region's potential for investment without mentioning names, contacts, or secret plans of potential investors. Further, consultants working on a commission basis should stick to real estate operations and divorce themselves from location studies.

Historical Events

At the time that Price Waterhouse took over Plant Location International in 1989, several historic events were taking place, each of which had a bearing on the investment climate and decisions in Europe.

The Unification of Germany

A few months after the sale of Plant Location International to Price Waterhouse, the Berlin Wall, which had separated West Germany from East Germany, was demolished. Suddenly, Germany became the largest and most powerful nation in Europe, and that event had a significant effect on the unification of Europe. The era of Chancellor Kohl set in motion a series of changes that had a significant effect on the evolution of investment in Central Europe.

Central Europe's Economic Development

Hungary, Poland, the Czech Republic, Slovakia, and some of the Baltic countries made the transition to a democratic lifestyle and looked to the West for economic support and investments. Most of these countries were on a list of applicants for entry into the European Union. In fact, on May 1, 2004, they became members of the European Union, increasing the number of member states from 15 to 25. The expansion of NATO's area of influence to the Russian border was underway. Fifty years of peace since the end of World War II and the end of the Cold War between East and West opened up new alleys for Europe's economic expansion.

Russia

The Soviet Union fell apart and the old federated states became independent. The situation returned to where it had been during the czar's rule at the beginning of the twentieth century. The clock had been set back.

The European Union

The European Union was developed starting with the advent of the Common Market, the opening of borders with the Schengen Agreement, and the Treaties of Amsterdam and Maastricht. Those developments changed the landscape of opportunities for site location.

The European Parliament

The European Parliament, headquartered in Brussels and with operations in Strasbourg, was one of a series of landmark institutional developments that caused unparalleled changes in the economic history of Europe. The European Commission's submitting to the control of the European Parliament, elections of a European Parliament, transition of the European Monetary Union and the birth of the euro— the common currency accepted by eleven of the fifteen European nations in the European Union—have all played a role in changing the economic landscape of Europe. *All these historic events have fundamentally changed location decisions.*

Despite all of these changes, one principle stays constant in site location: *the importance of "location, location, and location"* in the twenty-first century.

In this chapter, the main principles and practices involved in sound location studies have been defined. Now we turn our attention to analyzing each of the components thoroughly and, from experience, quote stories that demonstrate the advice contained in this book.

Project Requirements Questionnaire

I. Objectives

II. Site and Buildings

III. Financial Assumptions

IV. Production, Market, and Distribution

V. Raw Materials and Industrial Services

VI. Labor

VII. Utilities

I. OBJECTIVES

1. What kind of facility is to be located or relocated (offices, plant, research center, warehouse, etc.)?
2. What candidate countries and regions are being considered for the facility?
3. How does the facility fit in with your strategic business development plans?
4. What are your current thoughts about:
    a. Proximity to and links with existing facilities of the group?
    b. Need for financial incentives?
    c. Proximity to competitors?
    d. Type of investment?
      - Build
      - Purchase
      - Lease
      - Leaseback
      - Lease purchase
      - Advanced factory
    e. Advisability of joint ventures?

II. SITE AND BUILDINGS

1. Area and shape
    a. Minimum required area in acres or hectares
    b. Preferred shape of site
      - Rectangular   yes ( )   no ( )
      - Square   yes ( )   no ( )
    c. Special soil-bearing requirements
2. Special requirements
    a. Need for prestigious location

    b. Proximity to highway or town
    c. Proximity to airport
    d. Other requirements
  3. Transport facilities required
    a. Water transport   yes ( )   no ( )
    b. Railroad siding   yes ( )   no ( )
    c. Parking for cars and trucks
  4. Total covered area (in square feet or square meters)
    a. Initially
    b. Ultimately

III. Financial Assumptions

1. Projected cost of site and buildings
2. Estimated cost of equipment and tooling
3. Raw material and finished parts inventory
4. Other working capital requirements: cash, accounts receivable, work in process
5. What is presently contemplated for financing?
6. Value of projected annual sales initially and ultimately
7. Projected operating income before taxes

IV. Products, Market, and Distribution

1. List of products of the new plant
2. Geographic market area to be served by the new plant
3. How the products are moved to market
    a. Direct sales   ( )
    b. Distributors   ( )
4. Annual volume of products (weight and value)
5. What is the proposed distribution pattern by region and cities?

a. Truck   ( )
   b. Rail   ( )
   c. Air   ( )
6. Information about transport
   Description of the freight:
   a. Drums   ( )
   b. Cartons   ( )
   c. Wood cases   ( )
   e. Palletized units   ( )
   f. Bags   ( )
   Average weight, size, and value of one package unit.
   Special problems of loading and unloading.

V. Raw Materials and Industrial Services

1. What major raw material does the plant need to produce the projected volume of products?
2. Major present and potential sources of supply of principal raw materials
3. Volumes of raw materials
4. How are the raw materials transported?
   a. Truck
   b. Rail
   c. Waterway
   d. Air
   e. Tonnage or volume per shipment
   f. Special transport aspects
5. Requirements for small foundries, machine shops, and other local services

VI. Labor

1. How many employees and what types of skills are required for:

   a. Management

   b. Office

   c. Research

   d. Others

2. Enumerate the labor requirements of the new plant initially and ultimately, including number and types of skills required.

3. Are there any known plans for use of foreign employees? How many, by job?

4. Are there any special personnel problems that you anticipate as a result of previous experience?

5. What is management's attitude toward union and non-union workers?

6. What shift work is contemplated?

VII. Utilities (Initially and Ultimately)

1. Electric power needed for the plant

   a. Total installed power (KW or KVA)

   b. Estimated monthly consumption (KWh) and seasonal characteristics

   c. Daily period of consumption/peak loads

   d. What are the normal variations in your power consumption?

   e. What are the normal variations in your power consumption pattern?

2. Water needed for the new facility

   a. Volume of sanitary water, per day, and hourly peak loads

- b. Process water
  - Processes requiring water
  - Foreseeable monthly consumption
  - Can a portion of water by recycled? What percentage of total process water?
  - Time, duration, and characteristics at peak times
  - Special quality requirements (temperature, hardness, iron, etc.; sanitary water, canal water, or river water)
- c. Cooling water can be recycled
  - Hourly and daily consumption and eventually peak loads
  - Can a portion of the cooling water be recycled? What percentage of the total cooling water?
  - Special requirements (special inhibitors, etc.)

3. Waste disposal
   - a. Sanitary wastewater
     - Daily volume
     - Time, duration, and characteristics of peak loads
     - Does the company normally treat the sanitary wastewaters?
   - b. Process wastewater
     - Daily volume
     - Time, duration, and characteristics of peak loads' pH and temperature
     - Chemical analysis of waste stream(s) before and/or after own treatment
     - BOD-COD toxic elements, heavy metals, etc.

     Does the firm normally treat one or more waste streams?

     According to the company's experience, can the waste stream(s) be treated by a normal municipal treatment plant? What would be an ideal size and what process would be applied?

   c. Cooling
- Daily, monthly volumes
- Peak loads
- Outlet temperature
- Concentration of additives

   d. What other pollutants would the plant produce?

4. Heat requirements in BTU or Kcal per hour
   a. Process use
   b. Area heat
   c. What types of process heat require specific fuels (natural gas, propane, fuel oil, etc.)?
   d. Do you plan to install standby boilers using alternative types of fuels?

Chapter 2

The Project Definition: Basic Requirements

The first job of the location advisor is to understand the elements of a location decision that may lead to the success of the projected investment. The advisor must recognize the problems that may emerge from the choice of a particular site, and must understand the factors that affect the elements of the decision—factors that may result from the choice of a specific location on a particular site.

There is no need to delve into the secrets of the production processes, the quality of the product, or the management structure when making the assessment. The only elements to be gathered for the basic requirements are those that have a clear impact on the place of location or on the locational strategies of the industrial investment.

The project definition, therefore, is the unitary description of all factors affected by the siting. An example of the *locational facilities questionnaire* used for this purpose is shown at the end of the previous chapter. The questionnaire is brief and to the point: it requests information relative to the *objectives* the company wants to achieve in making the proposed investment, the *site* and *building* needs, and the *financial assumptions* used to determine:

- Investment in fixed assets
- Raw material costs and the nature of the raw materials
- Projection of annual sales
- Working capital requirements (including accounts receivables)
- Production processes
- Methods of financing and projected profit before taxes
- Market and distribution patterns
- Industrial services

- Labor (quantitative and qualitative)
- Utilities (power, gas, water, fuel, and ecological requirements)

A visit to a similar plant is very helpful, since the trained eye can catch many details that the company may consider obvious and therefore won't state when they complete the questionnaire. Careful observations will gather useful insights into the quality of labor, types of processes, conditions of work, male-female ratios, dexterity, and many other aspects of the plant's operation.

It is important to stress that the contact between the location advisor and the client be made by one person only. The location advisor appoints a project manager to work with the company. The company will indicate its permanent contact, which could be the facilities planner, the chief engineer, the production manager for the new plant, or the vice president for the geographic region. For smaller companies, the choice is likely to be the president, with the assistance of the treasurer.

The analysis of basic requirements should be the result of taking *five initial steps* prior to starting the locational investigation.

Step #1: Detect the Critical Location Factors

Among the project's basic requirements, a number of critical requirements will not be included. *A critical location factor is one without which the project cannot be successful, or without which the project cannot be accomplished.* A lot of skill is involved in finding these factors among the answers to the basic questionnaire.

A determination must be made of which factors could be "critical" under certain circumstances because of poor timing or large costs. For example, suppose that an investment requires a railroad siding for the transportation of its raw materials and/or outbound freight of finished products. If an economic analysis indicates that road transportation is too expensive, then rail transportation becomes preferable.

Many areas have railroad lines alongside the industrial sites. If the railroad is used by a fast, intercity, or express train, no siding would be allowed, since a rail siding would slow down rail traffic or potentially create a security problem. Also, the cost of a railroad siding can be a major consideration. A railroad company will agree to put in sidings only if the volume of freight transport is sufficient on an annual basis. If the cost of building a siding is too high, the investment may become prohibitive for both the company and the railroad.

Step #2: Select the Determining Factors for the Project's Location

Analysis of the basic requirements will quickly bring to light that some factors in the location analysis are more important than others. Also, it will show that one or more factors will play a determining role in the company's decision.

From experience, it can be said that while 60 percent of all location factors are similar or equal in importance, and therefore play the same overall role in decision making, 30 percent of the location factors can be tied to specific industry groups or special production processes. The other 10 percent of the location factors can be attributed to company policies or the attitude of management.

A typical instance can be found in the automotive industry. Take, for example, two location studies for an automotive plant that were conducted in the mid-1960s and early 1970s, one for General Motors and the other for Ford Motor Company. Both studies were directed to locating in Europe.

For General Motors, the initiative for the study starts with its Opel subsidiary in what was then West Germany. The basic requirements were spelled out in great detail. Intensive economic and engineering studies were conducted, for which the locational requirements could be identified. For Ford Motor Company, the initiative started in Cologne. The same details were provided to each company. Of all the relevant location factors and principles, 60 percent of the factors were applicable for nearly all the projects, and 30 percent were applicable to the automotive industry.

In the case of General Motors, recommendations provided by the consulting company or by the company's team were sent to European management. After further discussions and evaluations, followed by a reduction in the number of sites being considered (and after site visits), the final recommendation was given to the finance manager's team and then to the board for final approval. For the Ford Motor Company, the procedure was the same, except for the final step. Mr. Ford flew to Europe, met with statesmen, including the prime minister in some cases, and proceeded to select a location. He made the final decision, to the amazement of all, while the screening process was still underway. Fortunately, things have changed considerably at Ford since those days.

Doing it Yourself

"Once upon a time . . ."—the story could start that way—I met a lady and her husband whose mission was to locate a plant in Europe. Why were they living in Brussels? They were living there because the people of Belgium had been nice to her husband, an officer in the U.S. Army, who had been wounded in the Battle of the Bulge in World War II.

"Where to locate the plant in Belgium?" was the only question to be resolved. How would they determine the ideal location in Belgium? Two criteria had to be met: that it be located a maximum of a half-hour drive from their home in Waterloo (the place where Napoleon lost the war against the Allies in 1815), and that there be a local administration able to speak English (the couple disliked translators). The couple drove in all four directions from their home and ended up south of Brussels.

The location fit their requirements. It was only twenty-four minutes from their house. They met a town secretary who spoke excellent English. Further, they were able to locate on a large plot of land (wrongly zoned but not a problem, since everything could be fixed for a price) and at a decent cost. When I asked my usual questions about the availability of labor, professional capabilities, transport infrastructure (a highway with local access), and more, the couple waved them off as unimportant.

When I pointed out the problems relating to militant labor unions, the wide labor recruitment range for large quantities of manpower, and poor infrastructure, my concerns again were discounted. This company had not experienced these problems with its plants in mid-America, it had never had to deal with unions, and was accustomed to paying the highest wages in the area. Therefore, to the couple, the future seemed bright in Belgium.

Several years after the plant was built, I was invited to the farewell party of the boss and the arrival of his replacement. I faced the lady of the house with a smile. When she noticed me in the crowd, she came to me and said, "Professor, I should hate you. We had every problem you predicted. This choice cost us a lot of aggravation and money, but my main concern was the health and comfort of my husband."

I asked her whether they would act the same way if they had it to do over again. She replied, "No." She kept her word. Several years later, we were asked to do the location for their company for a plant in Turkey.

A powerful person on the board, such as a president, or an owner of a small- or medium-sized company tends to take the final step without waiting for the final analysis and recommendations of the people involved in the study. This tendency is the last "unforeseen" locational factor.

We recall a story concerning the period of the first electronics boom in Silicon Valley. Many companies, smaller than $100 million in sales, were restless and wanted to expand into Europe. At that time, European companies in the electronics field lagged behind Silicon Valley companies by ten years in research and development. When visiting a company president in Silicon Valley, we learned from him that an analysis of growth of a number of electronics companies indicated that several of them had developed markets in Europe to such an extent that more than half of their total sales were being exported to Europe. The shipments were being made by plane to satisfy the needs of American assemblers located in Europe. These same electronics companies were also supplying the United States with operations of client companies.

When we drew the attention of one company's president concerning the dangerous situation resulting from the high costs of supplying by plane, we asked, "What if Siemens, Philips, or other European companies start to make this kind of product locally in Europe at a lower cost?" The president answered, "You're d*** right!" With that remark, he turned around, grabbed the phone, and called his wife. He said, "Darling, get your suitcase ready. I have finally found a great excuse for taking that trip to Paris that we have talked about for such a long time!"

This president needed no further advice. He spoke with many European government representatives and was convinced that they had all the answers. He took action based on those conversations. We learned later that he had been talked into entering a joint venture with a major nationalized company. That company held all the control and learned everything there was to know about the American company. The president had turned over his know-how and experience in a venture in which he owned only 49 percent. Two years later, the local European company bought out that 49 percent for "a song."

The Most Important Factor in a Location Decision: A Woman

In the mid-1960s, a charming reporter in London interviewed me after the issuance of our report on the relocation of Covent Garden Market. She asked me, "What, in your opinion, is the most important factor in a location decision?" My reply: "Women." We had come across a number of cases in which a woman played a decisive role in the site decision.

Of our numerous stories, we might mention the "wife of the general manager," the "daughter of the vice president–international," the "mother-in-law of the project manager," etc. The following story is the most unbelievable.

One time we had an assignment with an American company to find locations for two plants: one in Belgium and one in the Netherlands. The product lines were entirely different but there was the same boss for both—a Dutchman. We recommended a site in Belgium for product X and one in the Netherlands for product Y.

When the decisions were announced, to our great astonishment, the plant proposed for Belgium went to the Netherlands, and vice versa. Many years later, when presenting the results of another study, we learned how that switch had been made. The company decided to put the Dutch manufacturing chief in charge of the plant of product X in Holland rather than in Belgium at the insistence of his Dutch wife, who did not want to live in Belgium.

Step #3: Beware of the Figures

For any study, one must be aware that the output is only as good as the input. Wrong input leads to wrong output, and therefore to wrong conclusions. The economics of a location study must be based on "reasonable assumptions," not on "exact figures." We have seldom been involved in a project where the data wasn't altered during the study or after it was completed. The order of magnitude of the numbers must be realistic, so the variations are not too great. Industry averages are less useful in studies, although they can be used if there is an absence of more precise information.

Another aspect in the use of numbers is the flexibility of figures and the ranges used. For example, the importance of utility costs is often shielded by irrelevant figures provided in the listing of basic requirements. A location study will therefore always make reference to the basic figures provided. Unit cost figures will be provided and will be used as a basis for recalculations at a later time, if necessary.

Many projects are insufficiently defined or studied at the time the location study is initiated. Delays in the project can be the result of changes in cost calculations or inaccurate forecasts of sales in the years following initial production. The resulting adverse effects are delays in the project and in cost overruns.

Step #4: Define the Nature of the Investment

The relevance of the location factors changes greatly when analyzing a capital-intensive investment, versus a labor-intensive investment project. A project is called *capital intensive* when the investment cost per job created is high. On the other hand, a project is called *labor intensive* when the investment cost per job is low. This definition, however, is not simply black or white.

A $200 million chemical plant that is expected to employ 100 workers with an investment of $2 million per worker is obviously a capital-intensive investment. An electronic assembly factory requiring a $10 million investment in fixed assets that employs 400 workers, mostly female, would require $25 thousand per job created. In the latter case, the project is obviously a labor-intensive project (compared to $2 million per worker for the capital-intensive project).

Many investments, however, are in the medium range. Therefore, it requires considerable skills and experience to classify the project and to define the nature of the investment.

Step #5: Look at the Intangibles

The quality of life, investment climate, degree of government intervention, and bureaucracy are the main concerns faced by potential investors in Europe or elsewhere. The list of intangible factors, those which cannot be expressed in dollars and cents, but nevertheless have an effect on investment costs, production costs, and profits, is growing steadily.

In the early 1950s it was easy to obtain permits and few objections were raised against a company choosing to establish plants all over Europe. Unemployment was high, purchasing power and labor costs were low, and consumer products were very much in demand. In the "golden 1960s" employment reached a peak and labor unions raised labor costs faster than productivity. Inflation, a concern to most governments, increased because of social programs and virtually unlimited borrowing. People had begun to live beyond their means. Finally, OPEC "threw a stone in the pool" by increasing the price of petroleum, which in turn resulted in higher energy costs. The result of these forces was more government intervention, more deficits, more spending, and more monetary problems.

Ever Heard of the Mafia in Reference to Location Decision?

Many years ago we were making a plant location study in southern Italy, when we came across the most remarkable situation that we are sure is a phenomenon of the past or else the situation has now taken new forms. In trying to study the importance of certain personalities in political, administrative, or union circles, we were frequently pointed to one specific name. It seemed that only he could give an answer to any question about the labor force, the cost of land, union activities, permits, the availability of sites, etc. Finally, with our Italian manager we decided to contact this individual, who happened to be a very highly positioned official in the regional city of Napoli (Naples).

We made an appointment through the proper channels and were informed that our request was accepted and that the meeting could take place at the government building. An armed guard guided us to the sixth floor. After twenty minutes waiting with the usual strong cup of coffee, we were introduced to the mysterious person. He received us with much charm, stretched out his hands, and thanked us for the interest we had shown in considering investment in his area. He knew all about the prospective investment. He gave us a complete picture of the economy of the area, the unemployment problems, the tranquility on which they insisted, and the investment climate he defended against the "conquerors" in Rome.

He proposed that we limit our search to certain towns and predetermined sites. We should not go around asking questions, since this could create suspicion and confusion among his citizens, fellow organizers, and owners. He said that he would provide all the answers we needed. When we asked how we could clarify his recommendations, he promised to send some person on the next day at 3 p.m. to the room at our hotel (he knew the name, the room number, and our dates of arrival and departure). He would then provide us with a map of the area. We were to look at the map, make notes, and then destroy the map, so it would not be taken out of the country.

> At exactly three o'clock, a messenger came to the room, asked to see our passports, checked our pictures, and delivered a roll with the map. That map contained all the information we needed. We took the necessary notes and then destroyed the map as we had promised. When we checked out the next day, the manager asked us for the pieces of the map. He went to our rooms to verify that nothing was missing, and only then did he let us pay our bill and leave.
>
> Was it Mafia interference or proof of good government? Either way, it was an interesting experience never duplicated in our forty-plus years of practice.

Intangibles have been considered in many ways. To assist the amateur location analyst, three groupings of intangible factors need to be considered. They are:

- The investment climate
- The ten main intangibles applicable in nearly all cases
- The Business Environment Risk Intelligence (BERI) rating

THE TEN MAIN INTANGIBLES

The investment climate is the result of a number of factors that determine the degree of success of industrial and commercial investments in a country.

The current situation, the degree of stability, and the trends for the decade must be analyzed in depth for the following key factors.

1. Political situation
2. Economic situation (expected growth and degree of inflation)
3. Monetary situation (evolution in exchange rates and restrictions)
4. Social legislation evolution (hiring and termination policies and cost)
5. Labor climate (management, union situation, strike history, trends in participation in management, and profit sharing)
6. Methods, availability, and cost of financing investments
7. Incentives available and depreciation policies
8. Taxation

9. Evolution of labor availability (industrial employment—unemployment rates, productivity, absenteeism, and labor turnover)
10. Market expansion and protectionist and nationalistic attitudes

THE BERI RATING

The BERI rating is one of the systems for business risk assessments used in evaluating the various countries attracting foreign direct investment (FDI).

The rating is based on a composite score derived from taking one-third of the operational risk index, the political risk index, and the remittance and repatriation factor. In the rating system, the higher the score, the lower the risk (see Table 2.1).

TABLE 2.1. BERI Ratings

Risk Level	Score
Low	70-100
Moderate	55-69
High	40-54
Prohibitive	0-39

Chapter 3

Critical Factors Analysis

THE CRITICAL FACTORS

In order to determine where *not* to look for sites, one must start to identify the critical factors that will exclude a number of areas if the study is for a location in Europe. The analysis of the basic requirements will indicate clearly where it will *not* be possible, feasible, or economically desirable to locate the plant.

What Is a Critical Location Factor?

Misconception abounds regarding the meaning of a critical location factor. Indeed, almost always a technical solution may be found for any problem, but what about the costs of achieving that solution? Remember, "any fool can succeed if he has all the money and time in the world." Let us look at a few examples of what went wrong as a result of neglecting one or more critical factors.

Example #1: A company producing piston rings and other motor components for large automotive companies wanted to build a plant in Europe to serve the American assembly companies operating there.

During this study, two factors emerged: one critical and another less critical, but still expensive. The parts were manufactured in inches and feet, measures used in the U.S. automotive industry. The machinery had already been ordered when it was discovered that the American automotive clients in Europe, including those in the United Kingdom, had already switched to the metric system. The second less critical factor was the absence of contracts in Europe. The U.S. automotive companies assumed that it would be "automatic" to deliver parts to the American clients that were operating in Europe.

Because the products were produced to American standards, i.e., inches and feet, the company had no choice but to ship them back to the United States for sale in this market. The extra costs involved in shipping the goods back drove the total cost of the products too high, and therefore rendered them uncompetitive. The plant was closed after two years of production. The losses amounted to about $10 million.

This company followed the example of the Monroe Auto Equipment Company, a U.S. producer of shock absorbers. Monroe had no customers in Europe when the plant was installed there, so the first two or three years' worth of production was shipped back to the United States. The production costs in Europe were low, however, so when the company developed European and American car assembly companies as customers, it became a profitable after-market business very quickly. Monroe is now part of Tenneco Corporation.

The critical factor for the first company described previously was that it had the *wrong product* for the market. It produced products to U.S. standards, and not to metric standards required in Europe.

Example #2: The president of an American company enjoyed the San Isidro festivities in Madrid in early May so much he decided to locate a new production facility near the city.

All went very smoothly at first. The local representative found the site. His brother, the architect, made drawings for the building. An uncle, who was a notary public, wrote the deed for the land. A cousin provided the permits. Another cousin arranged for local bank contacts.

There was just one hitch: the plant imported parts for the products daily from Barcelona to Madrid, a distance of 600 kilometers (380 miles). This situation occurred before the express highway was built to connect the two cities, so the roads were relatively poor. The Madrid area, where the final production took place, represented only five percent of sales. The rest of the production at the plant near Madrid was shipped to France, Germany, and the United Kingdom.

The critical factor in this case was *transportation cost.* The high transportation costs of parts going from Barcelona to Madrid, and the subsequent high transportation of getting the finished products from Madrid to the rest of the Spanish market and to export markets, combined to make the products significantly higher priced because of the choice of the location near Madrid.

Example #3: A large U.S. company intended to relocate a plant south of Madrid to an area north of the city.

Payment was made for an option on a piece of land north of the city. Unfortunately, the site was zoned incorrectly. The local authorities could issue a building permit, but the national planning department was strict in applying the zoning rules. No amount of money was able to convince those authorities to change their views. The result was a loss of time (almost two years) and money (the $50,000 payment for the option).

The critical factor in this case was the *issue of permits.*

Example #4: A plastic extrusion plant was to be located in the southern part of Italy. High-tension lines were located nearby and a transformer station was also located a short distance from the proposed site.

After more than one year of studies and negotiations, the "ready-to-go" signal was about to be given, when a red flag was raised because a local company went out of business. The company that went out of business needed a continuous supply of electrical power. The ideal arrangement for that company would have been a closed loop connection, whereby the plant could receive electrical current from two power supply stations, so that when a power failure occurred and interrupted power from one source, an automatic switchover would occur to the second supply source. A few bad storms and a rupture of a cable near that factory caused interruptions in the power supply and forced the company out of business.

Fortunately for the plastic plant project, the critical factor of *reliability of electric power* was identified sufficiently early so that it was possible to ensure that it would not suffer the same fate as the company that went out of business.

Example #5: When Bristol-Myers planned to locate a penicillin base plant in Europe, the focus of the investigation centered on a reliable supply of water, both in quality and quantity as the critical factors.

With Bristol-Myers's usual thoroughness and experience in its existing unit in Syracuse, New York, water-sourcing studies were conducted in various countries of Western Europe. The best solution for the penicillin plant was found near Sesse Latina, which is located between Rome and Naples, at the foot of the Apennine Mountains. That location guaranteed a steady source of water supply of the right quality.

*Example #6: Although our professional services covered some 1,500 studies worldwide, only in one case did we find what was undisputedly **the** site.*

This case involved a location study for an electric foundry for motor blocks for General Motors in Europe. We located a 500-acre, low-cost flat site, which was almost square, located on a canal, and nearly equidistant from its major assembly plants in Europe. This site also had excellent infrastructure, including good availability of gas, water, and power.

For the supply of power, a closed loop with *three* electric power stations could be provided which would ensure *absolute certainty of an uninterruptible power supply at all times at cost conditions that were unparalleled at that time.* Unfortunately, the energy crisis and rising costs put an end to that project, and it was never built.

The aforementioned examples are only a few of dozens of horror stories relating to insufficient attention to and understanding of critical location issues.

Determining the Area of Search

With the basic requirements analyzed and the investment project properly defined, we should be able to determine the area of search for the preferred sites.

The two steps for the geographic delineation were assessed, one negative and one positive.

- The negative step is based on the critical factors which make it impossible for one or more reasons to consider the region as a place for the investment.
- The positive step is based on the critical factors and the determining factors which could be physical or economical in nature.

Some Examples

A company producing packaged flakes is tied to a maximum distance from its customers because of the high volume, low weight, and low value nature of its products. Therefore, there is a limitation on the affordability of freight costs to get products to customers. To be able to compete with other products for the same market and maintain adequate profitability, the investment requires that the plant be located a maximum of 120 kilometers (80 miles) from the plant where the client companies are located. Those clients are capable of absorbing the total production of the packaged flakes plant. In this case, the total transportation cost is roughly 20 percent of the total product cost.

Similarly, it is generally accepted that cartons for packaging boxes, shipped in folded packages, must have a maximum delivery range of 160 kilometers (100 miles). At that point, shipping costs reach 50 percent of the product cost. Any shipment beyond 160 kilometers results in a loss.

Another example is the refining of white sand from the Mol area in Belgium. White sand is used in the ceramics industry in southern Italy. Ships of 15,000- to 20,000-ton capacity carry the unrefined sand from Antwerp to a port in southern Italy from which it is transshipped by truck to the sand processing plant within a short distance from the port of entry.

At this point, the total transportation cost, although kept to a minimum by efficiencies of shipment, is still far in excess of the cost of extracting sand at the site of origin. A complex process therefore generally affects the area of search. Besides the considerations listed above, there must also be a clear understanding of location factors at the regional and community level, as well as site specifications (explained later).

Remember, *there is no such thing as the "ideal" or "best" area in which to locate a plant.*

Local Recommendations

Locational recommendations are always the result of a compromise among factors. How then, can we determine the area of search? This can be done by elimination on the basis of an analysis of the location questionnaire and the critical factors. Where should the plant *not* be located? In other words, what "areas" (we are not yet talking about sites) should be eliminated from the search, and why?

Even if the client has stated a preference for certain regions, it is necessary to keep an open mind. Other regions may offer more opportunities to the client than the one he suggested.

Warning: Always offer four or five sites for consideration, not just one. Don't use the consultancy tricks that caused the disappearance of some advisors in the past. One of those tricks was for the consultant to concentrate on one or two areas that offer obvious advantages for the project, then fill up the report with "canned stuff," for example, secondary areas that were analyzed in previous studies.

The process of elimination and the resulting determination of the area and of the site search is one of the most critical phases in a location study. Here the factors *"experience, experience, and experience"* will determine the outcome of the study as well as the recommendations provided to the client. Be aware of the pitfalls one may encounter in conducting the location study, which include the effect of wining and dining by some regional development authorities, and unverified, biased information obtained from them.

Of course, it is the job of the regional development authorities to "sell" their area to new investors. *Don't accept offers to conduct regional studies if those for whom the study is being conducted have already decided to locate in a particular area.* Many consultants have fallen into this trap.

Never contact the authorities before analyzing the critical factors and the principal requirements defined by the client. *You owe the client total objectivity, confidentiality, reliability, faithfulness, and honesty. Any breach of these principles will result in the termination of your services.* Sometimes it is tough to turn down a good offer. Remember, however, that a satisfied client is your best salesman. (See "The Consultant's Role in Site Selection" appendix.)

The Physical Fit

We have now identified regions and communities which could be considered for the project being studied. The next phase includes establishing the "physical fit" of the project in various communities and the sites available in those communities.

From experience, we have learned that most location advisors go wrong in two situations.

- Their community selection is arbitrary and not based on the analysis described in Chapter 2.
- Their site selection is based on the physical characteristics of a piece of land without due consideration for the community in which the site is located. Engineering companies or people with engineering backgrounds usually make this mistake. We used to say, let us find the optimal regions and communities and we will produce the optimal site, even if we have to request and obtain the use of *compulsory purchase powers* by the national, regional, and local authorities. We would take these steps when we had established the basic needs for the investment project.

To compare the locational characteristics and rate the value of the community(ies) and subsequently the site(s), a general outline is used in which all aspects related to the area and site are viewed in a schematic. Needless to say, *not all elements are relevant for all projects*. The community analysis and site analysis will be dealt with in Chapter 8.

Regions, areas, and communities can be analyzed in the same way. Each region or community should take this analysis very seriously. They should have all the information necessary for potential investors that may consider the particular area for investment. If the industrial development for a community is to be successful, it must base its approach on the method used by the management organization considering the investment. This means that the description of the community must be based on the same questionnaire that is used as a model for investment projects by companies considering the investment, or by consultants who are acting on behalf of companies.

The authorities must understand which projects physically fit their communities. The method for doing that will be explained more fully in Chapter 8.

The degree of "fitness" can be established by means of a *computer program*. That program helps to calculate a score, i.e., the percentage of matching elements in the community. It evaluates the initial factors and features of great value to the project.

Before going more deeply into the various location costs, it should be pointed out that mini-locational studies remain limited to establishing weighed centers. (See the "The Language of Location" glossary.)

In evaluating transportation costs, both inbound and outbound freight, the optimal location is sited at the *weighted center,* that is, a geographical point where these costs are the lowest. *Inbound cost* refers to raw materials and other components necessary for manufacturing the final product. *Outbound cost* refers to the transportation cost of the final products.

It may be obvious that if transportation costs dominate the location decision, it is much easier and faster to develop computerized models in which optimization is achieved by calculating these costs for various assumed locations. While it would help to determine the optimal location, transportation costs are only one of many elements that go into making the final decision on a location.

Historically, Max Weber (see the "The Language of Location") developed the simplest approach to determine the optimal location a long time ago. He did so by combining an assessment of distance and weights of the products. The optimal location is the one with the *lowest transportation cost,* that is, the lowest weight multiplied by the distance.

A trial and error system can be computerized by combining a calculation of several optimal locations, then taking into consideration the size of the cost of weight for a number of locations. This assessment thus leads to the ultimate location decision on the basis of freight cost. The coordinates measured on a map will then be used to obtain the results.

When comparing the various locations (or cities) projected onto a map, the penalty can be calculated for each of the locations considered. The more sophisticated—or, shall we say realistic—approach is to project the location cost in the area of search.

Locational Costs

A company makes an investment in a production facility *to make money.* The client wants to know the order of magnitude of the differentials of the projected cost in the *proposed area of search* and on the *recommended sites*. Therefore, we must make an analysis of three groups of factors.

1. The nonrecurring cost factors at the time of the initial investment.
2. The recurring cost factors, or the repetitive costs, which are inherent in the production but will differ from area to area and from site to site. Unlike the nonrecurring cost factors, the recurring costs are more important in making the locational decision. These first two sets of factors are also called the *calculable* factors, since they can be expressed in dollars and cents.
3. The third set of factors is *noncalculable* and therefore called *intangible location* factors. Their impact on the project results must be evaluated and/or estimated. These estimations will be influenced by geography and time. This is probably the most difficult part of the study, subject to interpretation and subjective judgment, as we shall see in the section titled "The Weighing of Factors in Location Decision Making" on page 109.
4. The analysis would not be complete without considering the methods of financing investments, sourcing of raw materials, and the impact of ecology legislation.
5. The last elements to bear in mind are the community (or regional) and site analyses. Weights will be established that will consider all five elements, and that will result in conclusions and recommendations giving an overall view of the steps to be followed in implementing the decision. A major distinction must be made between labor- and capital-intensive investments.

In summary, we repeat that in the locational study, three groups of locational factors affect the final location decision of an industrial investment project. The first two groups of factors must be *calculated*, whereas the third group must be *evaluated* and *rated*.

The economic factors have a dual nature. The first aspect is the cost that occurs at the time of investment. We call these *nonrecurring* costs. These costs will have an effect on the location decision in only a few cases. Nonrecurring costs include site cost, building cost, and offsetting cash grants that lower investment cost. *Operational costs* are incurred constantly during production. These have a significant effect on product costs and therefore have a strong impact on a company's competitive position.

The third group of factors are the *intangibles,* which cannot be measured in dollars and cents, and therefore receive a rating.

Weighted Centers

Before going more deeply into the various location costs, it should be pointed out that some *mini-locational* studies are limited to *weighted centers.*

In terms of transportation costs, i.e., inbound and outbound freight, the optimal location is reached at the weighted center, a geographic point where these costs are lowest.

- *Inbound freight cost* refers to raw materials and other components necessary for the manufacturing of the final product that is shipped to the distribution center or the customer.
- *Outbound freight cost* refers to the transportation cost of the final product to its destination.

It is obvious that if the transportation costs dominate the location decision, the easiest and fastest way to get a solution is to develop computerized mathematical models that allow one to find the optimal location by calculating transportation costs for various assumed locations. While it is helpful to determine the optimal location, it is only one of many elements involved in making the final decision.

As said before, historically, Weber developed the simplest approach quite a long time ago. He determined the optimal location by combining distances and weights. The optimal location is the one with the lowest transportation costs, and also the lowest weight times distance.

A trial and error system can be computerized, whereby combinations of several optimal locations are calculated. This assessment will lead to the selection of the best location by taking into consideration the size of the weighted cost in a number of locations. The coordinates of those locations can be plotted onto a map, as a way to show the results.

When comparing the various locations (or cities) on a map, the penalty is calculated for each of the locations being considered. The more sophisticated, or realistic, approach is to calculate and compare the overall location costs in the areas of search.

LOCATIONAL FEASIBILITY STUDIES

If location of new investment plans is still in the early stage, on hold, or awaiting maturity, it would be advisable to make a *locational feasibility study*. Usually, when management gives the green light for a new project, they expect the project to be operational the next month. Hasty decisions, however, often turn out wrong and locational mistakes tend to be very expensive.

The locational feasibility study is an exercise performed to project a new production unit as an imaginary situation that can be transformed into reality as soon as the project is ready for decision making. The procedure is similar to the location study process outlined in this book. By providing data on a unit basis, it becomes easy to transform a feasibility study into a location study.

The main advantage of the feasibility study is to refine strategic locational policies and to project the search area that can be revised if necessary, thus eliminating pitfalls that would ultimately be found in the location study.

Furthermore, the feasibility study permits a range of assumptions that substantially reduces the time and cost of the ultimate location study. Setting a date for requirements is made flexible and any assumptions can be changed at a later stage when the final project is decided. The feasibility study is a substantial important time saver when a company finds itself slowly struggling with planning and strategy matters.

In the past, we have often created feasibility study outlines for large multinational companies. The frame of these studies does not differ very much from the location studies explained in this book.

MERGERS, JOINT VENTURES, AND ACQUISITIONS

Besides the development of greenfield sites, other ways to expand the business, make the company grow, and obtain further market penetration include mergers, acquisitions, and joint ventures. Almost all professions get involved in mergers and acquisitions (M&As), especially banks, law firms, accounting firms, and tax consulting firms. However, few realize the importance of both strategy and location.

1. Are the potential partners reliable?

To answer this question, one must understand the culture of doing business and making and respecting agreements. The era when certain companies in European countries had "three sets of books" is long past. However, this does not mean the figures they show are complete and properly valued. Therefore, make sure that interpretations are in line with current legal methods and procedures.

Of course, today the European Commission is imposing stricter rules, parallel to the ones practiced in the United States. At this point, however, Latin American and Asian companies have not followed suit.

2. Locations of manufacturing and other facilities

The locations of manufacturing and other facilities are also very important as potential partners may follow procedures to correct mistakes in past location policies. Also, trends in the customer's location, and thus the market, may have changed and therefore necessitated corrections, which are expected to be made by merger of companies and/or facilities.

3. Cases that illustrate pitfalls

Some of the stories reproduced in this book illustrate the pitfalls that are to be expected from unwise procedures in realizing mergers, acquisitions, and joint ventures.

Chapter 4

The Economic Analysis—Part I: Analysis of the Nonrecurring Cost Factors

The purpose of the economic analysis is to project the investment aspects of a number of sites and thus be able to compare the true economic values for each alternative. The economics deal only with location-related cost factors, or those costs that depend on and are the result of locating on a specific site. The objective is to provide cost figures that are as *realistic* as possible, not as *real* as possible. As the saying goes, "There are three ways of lying: one way is to distort the truth; the second is to withhold the truth; and the third is to use statistics." We refer here to *figures,* not *statistics.* Statistics are averages that are useless by the time they are available, for their values have changed since the statistics were gathered.

Information leading to realistic cost figures is obtained in three ways: stealing, borrowing, and buying. An example of using all three approaches can be found in the story "Steal, Borrow, or Buy."

The *second aspect* of the economic analysis is a comparison of the figures based on establishing *differentials.* Even better than such a comparison is the development of a *magnitude of differentials.* A difference of a few percentages in investment cost in dealing with a million dollar project is unimportant. For a half-billion dollar investment, however, a few percentage points can become a significant factor in making the decision.

"Steal, Borrow, or Buy"

It looks easy, but it's not. Up-to-date, honest facts and figures are hard to get. When clients ask us, "How can we get the right information?" our reply is invariably, "We steal, borrow, or buy." Once, we were faced with a rough study of retail facilities (multiple stores) in a British city. Our researchers spent several days trying to get information about the city. The officials were not helpful. They either had no information, were too lazy to collect it, or were leery about those "continental (i.e., European) fellows" spying on their town.

It was a fiasco, and we were about ready to throw in the towel. However, I had an idea. I called my assistant and gave him orders to fly to the city the next morning. Once there, he was to find out if a local university existed, call the university to see if it had a geography department, and then determine who the chairman of the department was. He was instructed to call this individual and arrange an appointment for me. Early in the afternoon, I received the call I was waiting for. "Yes, there is a university. Yes, they have a geography department and its chairman is . . ." I was astonished when I realized the man was an old friend of mine from my University of Chicago days (1947-1950). The meeting was arranged for the next day. It was a happy event. For two hours, we reviewed our lives. We had not seen each other for twenty-five years. Finally, Bill asked, "What brings you to this country?"

I explained my problem and we made a deal: he would arrange for me to meet the graduate students working on various aspects of the city and I, in turn, would give two lectures for the students. One lecture was on the unification of Europe (with which I had been very involved) and the other was on the location of industrial activities. We both kept our promises.

During my meetings with the students, I came across some very useful information. One student had written a master's thesis on the development of the city, and that study provided me with all the statistics I needed. He told me the number of pounds sterling that would be required to get his study typed and reproduced. I took out my checkbook and wrote him a check for the amount needed. He gave me a copy of the text, which we had typed in Brussels, on my return.

The second student was a cartographer. He had nearly twenty maps of the city with information on population density, land use and planning, shopping centers, industrial parks, telephones, income groups, etc. He needed money for professionally drafting the maps and for getting a half-dozen reproductions. Again, we took the maps with us and had them reproduced by a professional office.

A third student was interested in what we might refer to today as "mobility," as well as the social habits of the people in the city. We gleaned a lot from his research.

Instead of the proposed twelve weeks for completion of the study, we were able to present in eight weeks, to the surprise and delight of the client. When we made the presentation, the company's real estate team—which had been unable to carry out the assignment—entered

Analysis of the Nonrecurring Cost Factors 57

> the room where our maps were displayed; they could not believe their eyes.
>
> After the presentation, the CEO took me to lunch. Entering his office, he turned around and said, "Okay, now tell me, how did you do it?" My reply: "Mr. X, when I tried to sell you our services, you asked me how we collect information. I told you that we 'steal, borrow, or buy.' In this case, we did all three: we stole some, we borrowed a little, and we bought a little." "But where?" he asked. I told him about my University of Chicago buddy and he did not stop laughing for five full minutes.

The *third aspect* of the economic analysis is developing an analysis of trends. Indeed, as already mentioned, the time involved in making the location study, conducting the site search, and the start-up of production can be several years. Therefore, any figure quoted as "actual" during the study must be followed by a clear indication regarding the trends of the figures. The question must be raised, "Are costs likely to go up or are they sliding down?" One does not have to be a fortune-teller in order to give a clear-cut answer to this question.

The time involved in the length of projections is limited. In general, we have assumed in all calculations that the investment takes place in the year "zero" (the year production begins), followed by five operational years, by which time production will have reached maturity.

Projections beyond five years are merely speculation. In fact, the size of error is likely to increase from the second and third years onward.

Four main types of cost are incurred *before* production starts. These costs are incurred once and are not repeated. They may be considered "one-shot costs," also referred to as *nonrecurring costs*. These cost factors, however, are *in some cases* more important in the location decision than the *recurring cost factors*. This is the case with capital-intensive investments.

A $500 million refinery is capital intensive, since it employs few workers. The differentials discussed in the paragraph on *nonrecurring costs* may be more determinative than costs that recur annually (those costs will be discussed in the next chapter). In the case of labor-intensive investments, the overall labor costs will tend to dominate the locational decision.

CALCULATING INVESTMENT COSTS

Site Costs

Site cost is a better term than *land cost*. The latter refers to the raw land being considered for the project. The former refers to the land and all the elements of infrastructure needed for the project. A site will be a developed area of land. Site costs vary considerably, not only between countries, but also within various regions of a country. Further, costs will differ substantially closer to urban centers and to central locations.

We have included a table providing an illustration of the unit site cost differentials in various countries. However, the country averages—based on the current cost situation, as well as the size of the sites—differs so fast that such a table quickly becomes meaningless.

Development cost differences established by governmental authorities, or by private real estate developers, can be considerable and in some cases magnify the raw land cost. In West Germany, the investor's share of the development costs is paid at the time the deed for the land is signed at the office of the notary public. One pays the cost of the land to the landowner and the *entschliesungskosten* (development cost) to the authorities.

In Belgium, the government generally subsidizes the cost of infrastructure through the regional development organizations. Those organizations, in turn, pass the share of the infrastructure cost that is not subsidized on to the buyer. A few years ago the government of Belgium subsidized 25, 50, 75, or even 100 percent of the infrastructure cost as a way to reduce the cost of a site. We considered that a waste of taxpayers' money, for it has been proven that a low site cost by itself is seldom a major factor in making the location decision.

One way to keep site cost low is to avoid speculation. Land acquired by using compulsory purchasing power (as is possible in Belgium) and turned over at cost to individual companies later cannot be sold at a higher (speculative) price if that land is not used or is no longer needed for expansion. That land must be resold to the regional development authorities at the *initial cost plus the cost of the money invested at a reasonable annual interest based on current interest rates and the rate of inflation.*

Since the 1970s the number of greenfield sites selected and developed in Western Europe is much lower than in the three decades following World War II. Nevertheless, it is important to caution against certain pitfalls when considering site costs.

1. In recent decades, due to shortages of raw (undeveloped) land, more stringent zoning and planning regulations, and the growing influence of environmental protection agencies and Green parties, site costs have begun to skyrocket almost everywhere in Europe.
2. Quoted site cost must be defined properly by considering such factors as development costs (for infrastructure) and taxation. These factors can increase unit costs significantly. In Germany, the deed for land being purchased includes the *entschliesungskosten* (or infrastructure cost), which notaries pass on to the community responsible for paying for roads, sewage, street lighting, and other costs.
3. More and more expensive sites are being leased. In the United Kingdom, leasehold and freehold are old concepts and are widely used. In Continental Europe, the leasing formula is the result of the following considerations:
   - Investment in land cannot be depreciated. It is money tied up on the balance sheet without a return.
   - Especially when interest rates are high, companies reduce their borrowing.
   - Cash grants are seldom obtained for site costs.
   - By reducing the total investment, a better return is shown on the cash flow tables.
   - Leasing cost changes a nonrecurring, nondepreciable item into an annual operating cost, which then becomes tax deductible.

   While the site cost is very seldom a factor in the location decision (with the exception of distribution centers, which usually need to be located on a high-cost site, and therefore make leasing a primary consideration), for the authorities to "give away" the site is ridiculous, a waste of taxpayers' money, and totally ineffective in attracting the "right" industrial project.
4. In some countries, one can rent land, construct buildings on it, and then later acquire the land. He who owns the land can

become—under certain conditions—owner of the buildings. Ownership could result in a tax advantage.
5. When interest rates were high and money was expensive and scarce, such as in the 1967-1972 period, it was possible to have the local authorities buy (or offer) the land, build the buildings to a company's specification—which resulted in lower local property taxes (or total exemption)—and then sell the land and buildings at cost after fifteen years. The local government was not allowed to make a profit (e.g., Baxter Healthcare in Lessines, Belgium). In other cases, the regional development agencies (set up by the provincial government in Belgium) applied the same procedure (e.g., Mobil Corporation in the Virton, Luxembourg province in Belgium).
6. One must make sure that there are no hidden clauses requiring fees, commissions, and other gratuities for which *no tax deductions* can be claimed or, in some cases, no receipt is provided. Many such cases were found in Belgium, France, Italy, and Spain. Such fees can double the cost that was quoted initially.
7. Always check for hidden charges or mortgages that exist on sites being considered.
8. Another common trick is to buy an "option" valid during a period of one year, though one then runs the risk of not knowing if a building permit or an operating permit can be obtained. Several divisions of large U.S. corporations were the victims of such practices in Spain, especially in the early 1980s. The local authorities in the Madrid area (who were in on the deal) confirmed in writing that a *building permit* would be issued to the prospective buyer. It was not mentioned that the regional authorities had decided *not* to issue an *operating permit* that would allow for buildings to be used for commercial (or industrial) purposes.
9. Never take options on sites that need to be joined together. We saw several cases in Italy where three or more pieces of land needed to be assembled by land promoters. In spite of written guarantees, one of the owners of the land usually held out, demanding a very high price for his lot. That owner then split the additional profit with the other accomplices and the real estate agency involved. In one case, the Mafia interfered (positively) to eliminate such abuses.

10. Total site cost is generally only a small fraction of the real estate investment. However, it must be pointed out and stressed that the real estate lobbyists (agents as well as advisors) pretend that it is the most important factor in making the location decision. Further, the advice given by the so-called advisors and consultants generally are affected by the prospect of profitable commissions. *Warning:* Consultants whose real estate proposals are intermixed with location or other site selection work cannot be objective. Collusion and unethical practices are rather common.

The impact of zero site cost on the return of the project can be demonstrated. Remember, the only thing one gets for "nothing," is just that—nothing. You can rest assured that if a site is offered for free, most likely some deficiency exists that will result in a higher than normal development or building cost.

A Mother-in-Law Story

In the location decision concerning a small pharmaceutical plant, an analysis was prepared comparing the Paris region with Normandy.

The conclusions of the study were very clear: the Normandy region was the better choice. Within that region, we studied a number of sites, and the best site would have been in Caen.

During the presentation to the company's officers in the United States, the French manager indicated that our site search had been incomplete. He said that he had come across a site that was excellent from both the location and cost points of view. With that news, I contacted our company representatives in Paris with the promise that I would report the results the next day.

We discovered that the French manager's mother-in-law owned that site. She had been unable to sell the site because it had some flaws that would have made construction costs too high.

We never came across that French manager again.

Building Costs

1. It is vital to hire the right architecture and/or engineering firms when considering expanding operations into Europe. In some countries, e.g., Belgium, architecture and engineering compa-

nies are separate disciplines. During recent decades, engineering companies with architecture capabilities have expanded substantially. Most of the large American companies are already established in Europe, or have acquired European companies.
2. One warning must be considered. We have seen abuses made by engineering companies that hired a site location specialist (who was, in fact, a real estate agent working on a commission basis) in order to be sure that client would not deal with competitors.
3. Another pitfall is the use of a combined engineering-architecture construction company (the so-called total-package deal). That combination of capabilities results in less than the sum of the parts.
4. Wrong interpretation of planning and zoning regulations can result in expensive mistakes.
5. When working with permits, wait until all the paperwork is done and the regulations are clear. By waiting, you will avoid delays, expensive changes, and unexpected or unbudgeted investments.
6. Read the fine print on the construction contract, especially clauses that deal with revision of costs. Construction companies include labor cost increases due to inflation, governmental regulations, or negotiations between unions and the building industry in their contracts.
7. Look closely at the payment schedule and the conditions of payment. Determine when and how payments must be paid during the execution of the contract.
8. Avoid drastic changes in plans if that can be avoided. Additional costs resulting from changes in the plan will just provide builders with a higher profit.

Since each investment project includes a building of a different type and size, the following methods for determining building cost are commonly used.

In one case, the proper description of the buildings provided in the basic requirements of an investment project is submitted to architects, engineers, and/or construction companies for a *cost estimate*. Alternatively, the standard type described previously is used for making price comparisons in various countries or regions. In many parts of Europe there are building cost indexes, showing regional differentials. They are not generally available, however, so it is not possible to make comparisons between countries.

A more frequent assumption that is used is the construction labor cost. One can assume that the labor cost amounts to 20 percent of the total cost of the building. If we assume that the remaining 80 percent (comprised of material and other parts of the cost) is equal in all regions, the labor cost differential—thus, the magnitude of the differential—will fluctuate according to the construction labor cost.

Building Construction—Story #1

After locating a plant for a major U.S. pharmaceutical company, we were asked to advise on the selection of building companies that would be asked to tender for the construction. We provided the names of five large builders, knowing that each were capable of doing a good job. In the study, we estimated the total cost for the building to be 2.5 million euro (equivalent of the currency used at that time, prior to the establishment of the euro).

When the five cost proposals were submitted, we received a call from the country manager who was in charge of the project, claiming that our estimate of 2.5 million euro was too low by about 10 to 15 percent. The quotes he had received ranged between 2.75 and 3.1 million euro. Since it was an Easter Friday and a long weekend, I requested that I be able to study the five proposals and report to the company by midweek.

What we suspected turned out to be the truth. The lowest bid was 2.85 million euro; the highest, 3.25 million euro. The difference between the offers was close to 400,000 euro. On Tuesday, I called the number five builder and asked him if he was interested in quoting for a job. He declined, saying they were too busy at the moment. I repeated the same request with builders four, three, and two. All of them were uninterested for one reason or another.

Finally, on Thursday, I went to see the lowest bidder and discussed why our suggested price of 2.5 million euro was so much lower than his bid. He confided that he needed the job and had called his colleagues and offered them a "piece of the cake" if they agreed to inflate their bid quotes. That is, he would give them a share of the total price he was bidding if they let him have the job. To get off the hook, this builder offered to study the specifications again and prepare a new bid within eight days. This time he made a much lower bid, and the contract was signed for 1.8 million euro, thus providing substantial savings for our client. Because our first job for this client was so successful, we received eighteen additional assignments from the company in the following years.

Building Construction—Story #2

This second story runs parallel to the first story about building construction, but it does not have the same ending. In this case, the real estate manager of the company sent from the United States to supervise bidding and construction was taken in by the highest bidder, supposedly because that company would offer the best quality.

Later, we learned that he had been invited on the yacht of the owner of the construction company to sail for one week in the Mediterranean. We had no proof that there had been wrongdoing, though we had strong suspicions. Several years later, one of the sons of the construction company's owner admitted that a deal had been struck with the real estate man, who needed to take care of some personal debts in the United States.

Someone in the company must have "smelled a rat" because the chap was advised to leave the company. We have met that man at conferences many times since that episode, but we never got a study assignment from any company for which he worked.

Equipment Costs

1. In order to calculate the total equipment cost, one must consider four aspects:
   - If the equipment is ordered in the United States, the American companies are accustomed to dealing with the item and therefore no further explanation is necessary.
   - The equipment bought overseas in Europe or elsewhere (but outside the United States), where local procedures and customs play a role in pricing the equipment.
   - The transportation cost from the supplier (either in the United States or elsewhere) which the company can handle without undue difficulties.
   - The installation cost. Usually this is the most uncertain of the four aspects and, if no measuring or calculating is done, a rule of thumb calculating 10 to 15 percent on top of the cost of equipment should be used.
2. A general remark that should be made and applied to the analysis of both the recurring and nonrecurring cost factors is that in locational studies, one always refers to *order of magnitude* in dealing with figures. Indeed, it is impossible to develop "abso-

lute" figures, because the relative importance of the figures used will depend on the following:
- The place in which the site is located
- The time of data collection
- The origin of the figures

Avoid the use of statistics, for they are either out of date, not comparable, erroneous, tendentious, politically adapted, or specifically set up for a purpose, as well as subject to interpretation.

The equipment cost assumptions provided in the basic requirements of an investment project can be considered the same for all preferred sites, since the origin of the equipment remains the same wherever the plant is built. However, one can assume a labor installation cost of 10 percent of the total equipment cost. That amount will fluctuate depending on the country and region where the plant construction will take place.

The labor cost indexes of the regions are used. The relative importance in the use of figures in location studies should especially be considered when making location decisions.

Incentives: Cash Grants

1. The reasons for offering incentives are dual:
   - To attract new investments to create new jobs, especially in areas with declining economic activity.
   - To compensate the investors for problems, inefficiencies, and costs due to special conditions in the area. The greater the problems, the higher the need for incentives.
2. Forty years of experience have proven that no company will locate a new investment on the basis of incentives only.

    If incentives are offered in economically justified locations, one should take those incentives but be aware of the negative implications as well as the "goodies." Companies that want to invest the least possible amount of money will seek cash grants. Experience has proven that companies that chase maximum cash incentives do not last as long as companies that consider cash incentives as an extra factor.

    Furthermore, high technology companies and those with sensitive defense projects will avoid incentive applications as they are a nuisance. The government will impose so much pa-

perwork that the timing and cost involved will take a bite out of the cash incentive. Worse, however, is that information requested by bureaucrats will be passed on (or sold) to competitors, especially if the American company's new investment is considered a threat to local companies (no names, please!).
3. We have always been against giving and receiving incentives, especially cash grants. *In life, one gets nothing for nothing.* There is no such thing as a "free meal." Further, cash grants distort the competition among the regions of Europe that are trying to attract new investments to create jobs, stimulate the economy, or increase the population's standard of living.
4. Cash grants are a one-time shot. The economic disadvantages of a location on a given site could wipe out the advantages of a cash grant in a very short time, thus turning the location into a disadvantage.
5. Cash grants take the form of a sum of money paid out *during* the construction of the plant, prior to operations. It has always been our belief that companies should be rewarded for results, not for speculation.
6. One-time grants can also be given in the form of loans at reduced interest rates, or grants as a part of the interest on borrowed money. However, the "snake" could be fiscal, for in some cases these grants on interest are to be written off over a period of years, and not necessarily deductible immediately. In that case, a 15 percent grant could in reality be worth less than 9 percent.
7. The impact of the cash grant on cash flow tables can be important; however, the "small print" of the governmental regulations and the frequent changes of these regulations must be examined carefully. Furthermore, the European Union, aware of the competitive aspects of cash grants (and other incentives) is monitoring the application of various government programs very closely.

We now have the four nonrecurring location factors ready for the preparation of cash flow tables.

In one case involving a feasibility study for a chemical plant, we calculated the total investment cost differentials in some fifteen countries. The magnitude of the differentials of the total investment cost for a 16,600,000 (expressed in pounds sterling) building if built in the United Kingdom is rather considerable. The lowest total investment was projected to be in Portugal (19.3 percent less), a total of 3,226,000 less, followed by Greece (17.19 percent less), a total difference of 2,858,000 less. The higher investment cost was in West Germany (5.4 percent more), a total difference of 898,000 more. Was this project realized in Portugal or Greece? Neither; it remained in the United Kingdom because other more important factors (e.g., political and labor union pressures) were taken into consideration.

Chapter 5

The Economic Analysis—Part II: Analysis of the Recurring Cost Factors

While the nonrecurring or investment cost factors are important in making the locational decision, the recurring cost factors are—in most cases—the *determining factors* since they determine how the optimal profit can be made. Let us recall the three basic desires of the investor.

1. To make the lowest possible investment (when using its own money). This is determined by nonrecurring cost factors.
2. That the highest profit should be the result of the lowest operational or recurring costs.
3. That the least possible problems affecting projected investments arise. (These are analyzed in Chapter 6.)

LABOR COSTS

In *labor-intensive investment projects,* labor cost is *the* most important factor in making the decision. However, much confusion surrounds the definition of "labor costs." To the *employee* it is the net amount of money that he or she receives every month (besides the "sweeteners" or extras such as company cars, extra legal pension contributions, and health insurance for the family and others).

To the *employer,* labor cost means the total cost of an employee, that is, the *gross disbursement* resulting from the employment contract. This includes not only the gross sum agreed upon with the employee, but also the social security charges, "sweeteners" mentioned previously, contributions to numerous agencies (national and/or regional), employment tax (believe it or not), security arrangements (e.g., clothing), meals, and other allowances. The total cost of these extras can amount to 40 to 90 percent of the gross amount. It must be warned, however, that the high labor cost in a country of the European Union does not necessarily mean that this country is not to be considered as a favorable location when making the decision.

It is generally known that of the fifteen countries in the European Union, Germany and Belgium have the highest labor costs. Nevertheless, new American investments favor Germany (second after the United Kingdom), as well as Belgium. Labor efficiency, labor-union cooperation, and the favorable investment climate largely compensate for the high labor costs in these two countries.

It is pointed out several times in this book that a location decision is not the result of one predominant factor; rather, it is a compromise of all three groups of factors. Labor cost, usually a "hot potato," is generally blamed for the lack of new investments, but that issue is judged incorrectly. The highest standard of living occurs in the countries where the population has the highest net income, the emphasis being on *net.* For example, the labor cost in Belgium is high, but not prohibitively high. The money that the employer receives is too low. Inefficient and wasteful government, poor management, and political problems are to blame in Belgium, not the high labor cost. Comparative tables can demonstrate the impact of labor cost on labor- and capital-intensive investments.

TRANSPORTATION COSTS

Transportation cost is possibly the most neglected cost factor considered in the location decision. The *competitive position* of a product depends not only on the *product cost* when the product leaves the factory. One must make a distinction among three components of the transportation cost: (1) inbound freight costs; (2) outbound freight costs; and (3) the interplant freight cost. The best example of the importance of the three components of transportation cost is illustrated in the study of automotive assembly plants.

Inbound Freight Costs

The cost of raw materials is *not* a factor in the location decision. The cost of transporting those raw materials to the plant, however, can be very important. Transportation costs by rail, road, water (canal), or pipeline are very different and relate directly to the type of raw material or semi-finished product being transported.

Transportation cost is a major consideration when dealing with a chemical plant (pipeline is preferred), a consumer goods plant (delivered by road transport), or a bulk user (transported by rail). The railway companies in Europe—mostly nationalized units being privatized—have neglected chances offered in the 1960s and 1970s to offer lower flexible rates and fast deliveries. Currently the main highways of Europe are gradually becoming clogged, thus increasing transport times and costs. Those railroads must reconsider their practices urgently and study their roles in providing inbound, outbound, and in-plant freight.

Outbound Freight

In recent decades, a gradual evolution has taken place in the factor of outbound freight. It is now classified under *distribution cost* and includes the warehousing cost incurred before the product reaches the final customers. The study of distribution cost is gradually becoming an important new element in the location decision, particularly in low-cost consumer products and products consumed on a daily basis.

Interplant Freight

Interplant freight is only pertinent in an assembly plant, such as those in the automotive industry (e.g., Toyota). To reduce the interplant freight decades ago, the Japanese introduced the *just-in-time delivery* based on having suppliers located close to the assembly plant. The relatively short distances traveled in Japan made that method possible.

Further, the elimination of storage space for parts, coordinated deliveries with daily production, and provision of quality control at the suppliers' plants by Toyota personnel, all contributed to a just-in-time delivery system. This system not only reduced interplant freight cost but also saved space and improved financing costs of the suppliers (i.e., their need to finance what had been a certain inventory of parts that wouldn't be needed for days, weeks, or even months).

"Logistics" Is the Magic Word

Several mathematical (read: impractical) studies have been conducted to determine the impact logistics costs have on final product cost and therefore on the competitive position of the product. Conclusive results found that all three aspects of transportation cost should be considered when making a location decision. However, the decision also depends on the type of product, the market, distribution, distance of deliveries, the means of transportation used, the transportation infrastructure available, and the European Union policies that are gradually replacing or supplementing national practices.

UTILITY COST

Utility cost has never been a decisive cost factor in the location decision. This does not mean, however, that it should be excluded from the location study. Indeed, economies of scale can be achieved in dealing with large suppliers of utilities: power, gas, water, and pollution treatment.

Power Cost

While a location decision has never been based exclusively on power costs, power suppliers become a topic of discussion when dealing with large capital investment projects, such as those in the chemical industry. The power companies in the fifteen countries of the European Union have, for a long time, abused their virtual monopoly in their respective countries by charging abominably high rates.

The European Commission has decided to free the power market from those monopolies by opening the borders and encouraging competition. This will not be an easy task, since the politicians (local and national) have been accomplices in maintaining artificially high rates. These high rates lead to high profits and benefits to the authorities (local and national) who are partners in the distribution of electricity.

Some large chemical companies in the port of Antwerp plant are considering having power stations built on their main product sites, using one or more supranational utility companies. In some European companies (those in France, for example) it will be difficult to break the monopoly of the nationalized Électricité de France (EDF) and the waste of resources for the economy that has resulted from their position.

Gas

The remarks made concerning power can also be applied to gas suppliers. Further, the geographic distribution of available natural gas covers areas in Holland, the North Sea region (particularly Norway), Russia, and Algeria. Their supply is influenced by the geopolitical situation. Availability of gas from those countries is subject to fluctuations in time and geography. The chemical industry watches this situation carefully because of the importance of gas, not only as a source of energy, but also as a raw material for the petrochemical industry.

Water

The cost of water to both the consumer and industry has increased enormously in recent decades, not only because of limitations in the supply, but also because of the treatment cost and legal requirements necessary to protect the health of the population.

Pollution Treatment

The treatment cost of air and water pollution has increased substantially as a result of stringent requirements. These requirements resulted from legislation imposed in the European countries under the political pressure of the Green parties. Pollution treatment is *the only locational factor* that must be classified in all three groups of location factors.

- It is an investment cost (the installation of a treatment plant for the production unit).
- It is also an operational cost factor since the products used in water and air treatment will depend on the production process.
- It is an intangible factor because the magnitude of the costs of the two previous factors will depend upon legislation, which varies from country to country as a result of political policing.

Water and Waste

Water and waste have always been high on the list of determining factors in selecting a plant location, particularly in the chemical and pharmaceutical industries, but also in such large plants as automotive assembly lines. However, with the increased influence of the Green political party, legislation, requirements, and cost have increased dramatically over the past two decades.

The most difficult projects we have dealt with in terms of water and waste were in Italy for an antibiotics plant and in the United States for an artificial sweetener plant. Both projects needed massive and steady amounts of pure water. The first plant had a deoxidized stream, which smelled very badly. The second produced a large amount of wastewater that was very expensive and difficult to treat.

In the case of the Italian plant, the geographical pattern of the region and the all-important agricultural union were used to determine location. Many irrigation canals south of Naples were dredged during the early years of the Mussolini regime to be used for soldiers in the Pontini marches that settled people from the north. Those living in the canal region became victims of the inundations, and in many cases lost their homes and belongings. As a result of the dredge, the region became an uninhabited yet fertile agricultural area with high-yield crops. When a hydraulic study indicated the presence of large quantities of clear water from the Apennines, the only drawback was the cost of treating deoxidized water that had a bad smell.

Discussions with agricultural lobbies led to an agreement to use the wastewater as irrigation water and to pump that water onto the land. The results were spectacular. On the negative side of this action, the area could be spotted from a long distance because of the foul smell, particularly in the hot summers. On the positive side, the wastewater acted as a fertilizer, which increased the crop yield substantially.

Some time later, a treatment plant was built and the previous system was taken out of service because of ecological legislation.

TAXATION

In a locational study, nothing is more difficult than the calculation of taxes. Taxes are raised on all three levels: national, regional, and local. Companies are less concerned with regional and local taxes, since tax concessions are usually available to reduce their impact if the community is desirous of attracting new investments and of creating jobs.

There is a difference between the tax rates imposed (the statutory tax rates) and the rates actually paid by the company (called the *effective tax rates*). In research undertaken by the University of Maastricht (Netherlands) in 1999 for the Ministry of Finance in the Netherlands, a number of companies within the fifteen European Union countries were studied in terms of tax rates during the 1990-1996 period. The conclusions are most revealing. Table 5.1 shows the statutory and effective tax rates of all fifteen countries.

We can draw the following conclusions regarding the statutory tax rates shown on the left side of Table 5.1.

1. The average rate in the European Union is 36.45 percent.
2. Some ten countries show a lower statutory tax rate than the European Union average: Ireland (21.94 percent); Sweden (28.54 percent); Greece (32.53 percent); United Kingdom (33.35 percent); Finland (34.02 percent); France (34.70 percent); The Netherlands (35.00 percent); Spain (35.30 percent); Denmark (35.78 percent); and Austria (36.02 percent).
3. Five countries have an average rate higher than the overall European Union rate: Portugal (39.59 percent); Luxembourg (39.40 percent); Belgium (40.28 percent); Germany (50.05 percent); and Italy (50.48 percent).

On the right side of the table we see the effective tax rate in the European Union countries. This perspective is very different from looking at the statutory tax rates.

1. The average effective tax rate is 26.86 percent for the European Union.
2. Six countries show an average lower than the European Union effective tax rate average: Ireland (13.86 percent); Portugal (17.19 percent); Austria (17.67 percent); Greece (20.85 percent); Belgium (20.99 percent); and Spain (24.11 percent).
3. The other nine European Union countries show a higher average than the European Union effective tax rate average: Sweden (27.47 percent); United Kingdom (29.00 percent); Denmark (29.40 percent); Finland (29.82 percent); The Netherlands (31.80 percent); France (32.82 percent); and Germany (38.53 percent).

TABLE 5.1. European Union Member States Ranked by Statutory Tax Rates and Effective Tax Rates

Rank	Country	Statutory Tax Rate (%)	Rank	Country	Effective Tax Rate (%)
1	Italy	50.48	1	Germany	38.53
2	Germany	50.05	2	Italy	35.32
3	Belgium	40.28	3	Luxembourg	34.09
4	Luxembourg	39.40	4	France	32.82
5	Portugal	39.29	5	The Netherlands	31.80
6	Austria	36.02	6	Finland	29.82
7	Denmark	35.78	7	Denmark	29.40
8	Spain	35.30	8	United Kingdom	29.00
9	The Netherlands	35.00	9	Sweden	27.47
10	France	34.70	10	Spain	24.11
11	Finland	34.02	11	Belgium	20.99
12	United Kingdom	33.35	12	Greece	20.85
13	Greece	32.53	13	Austria	17.67
14	Sweden	28.54	14	Portugal	17.19
15	Ireland	21.94	15	Ireland	13.86

Source: Corporate Effective Tax Rates in the European Union. Faculty of Economics and Business Administration, University of Maastricht, April 1999.

Again, however, figures do not always show the true picture, especially in the case of multinational companies that have the possibility to practice intercompany or intercountry pricing and thereby reduce profits or report a higher profit in lower tax rate countries. We generally leave it to the tax professionals to deal with this item.

The practice of negotiated tax rates over a maximum number of years has become standard practice and is often stressed by large tax consultancies. It is important, however, to look at the reasons for large discrepancies between the statutory and effective tax rates. Different types of tax incentives offered tend to affect those discrepancies.

TAX INCENTIVES

1. The impact of tax incentives on statutory tax rates is given in Table 5.2. We can make certain observations.
   - The average European Union difference between the statutory tax rate and effective tax rate is 9.58 percent.
   - The *lowest differentials* are in Sweden (1.07 percent); France (1.88 percent); The Netherlands (3.20 percent); Finland (4.20 percent); United Kingdom (4.35 percent); Luxembourg (5.31 percent); Denmark (6.38 percent); and Ireland (8.08 percent).
   - The *highest differentials,* resulting from the highest tax incentives, are in Spain (11.19 percent); Germany (11.52 percent); Greece (11.68 percent); Italy (15.16 percent); Austria (18.35 percent); Belgium (19.29 percent); and Portugal (22.10 percent).
2. The statutory corporation tax rates, including local taxes, in the European Union member states are shown in Table 5.1. On average, the statutory tax rate is approximately 30 percent. The three highest rates are in Italy, Germany, and Belgium, and the lowest in Greece, Sweden, and Ireland.
3. All the countries provide tax incentives except Sweden, which has no tax incentives at all. The University of Maastricht study referred to above lists the following tax incentives offered in the EU countries.
   - Losses (by means of carry backs and carry forwards)
   - Fiscal unity (by compensating losses and profits between companies)
   - Foreign tax relief
   - Investments
   - Research and development (R&D)
   - Tax-free reserves (like reserves for probable losses)
   - Deviating tax rates for capital gains
   - Size of the company (especially small- and medium-sized companies)
   - Regional location (notably less-developed areas)
   - Cooperation, headquarters, and distribution centers
   - Foreign or exporting income (taxes at differential rates)
   - Employment (e.g., tax incentives related to the number of employees)

- Training of personnel
- Assets (exceptional depreciation for certain classes of assets)
- Industry (preferential treatment for certain industries)
- Age of company (deviating tax rates and/or rules for new companies)
- Foreign investors (like lower income taxes for expatriates)
- Leverage (paid interest being tax deductible, which is not the case for dividends)

The complexity of the value of incentives is especially demonstrated by the tax application of cash grants. Table 5.3, applied during the 1980s and 1990s in ten territorial units, shows the difference of this value. The table summarizes the application of incentives in a number of countries or city states, such as Berlin, whether the grant is taxable or nontaxable and depreciable or nondepreciable. A 25 percent cash grant, when taxable only resulted in a 16 percent real value. Of course since we drew attention to such anomalies, the administrations of the various countries have since introduced corrections. It remains that our remarks are still valid.

TABLE 5.2. European Union Member States Ranked by the Difference between the Statutory Tax Rates and Effective Tax Rates

Country	Amount	Unit
Belgium	40.3399	Belgian francs
Germany	1.95583	German marks
Spain	166.386	Spanish pesetas
France	6.55957	French francs
Ireland	0.787564	Irish pounds
Italy	1936.27	Italian lira
Luxembourg	40.3399	Luxembourg francs
The Netherlands	2.20371	Dutch guilders
Austria	13.7603	Austrian schillings
Portugal	200.482	Portuguese escudos
Finland	5.94573	Finnish marks

TABLE 5.3. Fiscal and Accounting Approach to the Cash Grant

Country	Non-Taxable	Taxable	Non-Depreciable	Depreciable
Belgium	X		X	
Denmark		X		X
France		X		X
Germany				
Zulage	X			X
Zuschuss	X		X	
Italy	X		X	
Luxenburg	X			X
Netherlands				
IPR	X		X	
WIR	X		X	
Republic of Ireland	X			X
Equipment	X			X
Buildings	X		X	
Northern Ireland and United Kingdom	X			X
Berlin	X			X

Chapter 6

Analysis of Intangible Factors

In the course of more than four decades of industrial location studies, we have dealt with some twenty-five intangible factors, along with others of lesser importance. These intangible factors can be grouped into four categories, namely:

- The investment climate
- The social climate
- The political situation
- The economic situation

While it is not possible to fully describe and evaluate all these factors, this chapter will distinguish the most important elements included in these four categories. Although the European Union tends to generalize many of these factors, more and more regional differences are being accentuated, shifting from national to local or even between local levels.

THE INVESTMENT CLIMATE

"To improve the investment climate" and "to create a favorable investment climate" are the two slogans one hears from regions and nations that want to attract investments. What is the significance of these two slogans?

As we stated earlier, *the industrialist hates trouble, dislikes problems, and likes to evaluate what could have an adverse effect on the planned investment.* The questions asked by the industrialist include:

1. Is foreign investment wanted or desired in the region or country?
2. What is the attitude of the political leaders and the population toward capital imported from abroad?
3. Will the host regions or nations question the value of decisions made in a foreign country?
4. Under what terms and conditions are incentives offered?
5. Are foreign investors treated in the same way as local investors?
6. Are the new, foreign companies encouraged to participate in the life of the locals, or might they be repulsed and excluded?
7. How are foreign managers viewed?
8. Can the company hope to receive fair treatment in the application of the laws and in obtaining permits?
9. Will the company be allowed to contribute to external plant activities and other festivities?
10. If the company personnel don't speak the local language, will that factor harm their relationship with the local people?

These are the ten questions most frequently asked by companies *that want to be good citizens in foreign countries.* The replies to these and other questions constitute the climate needed to attract investment.

THE SOCIAL CLIMATE

It is vital that management and the union representing the workers get along. Many years ago a large U.S. company from the Midwest asked us to find a location where they would not have to deal with a union. We told them that if they wanted to have labor peace and good union-management relations, they should talk to the representative unions before deciding on a location.

The local politicians in Europe assured the American company it would have nothing to fear from the unions. Thus, the company ignored our warning. Before its new plant was in operation and the first 200 workers employed, the company encountered several costly strikes. The company learned its lesson from that bad experience and hired a social consultant to help effectively deal with the local unions.

On the labor front, the situation has changed substantially during recent decades in two principal ways.

1. The European Commission created the office of Directorate General for Employment, Industrial Relations, and Social Affairs. The annual publication *Employment and Labor Market in Europe* summarized the policies imposed (or proposed) by the Commission and the employment results. The report shows:
   - The key employment indicators in the European Union (E15) comparing 1975, 1985, and 1990 (each year) to 1996
   - The key employment indicators for the same eight years in each of the fifteen countries: Belgium, Denmark, Germany, Greece, Spain, Ireland, Italy, Luxembourg, The Netherlands, Austria, Portugal, Finland, Sweden, France, and the United Kingdom
   - The macroeconomic indicators'—output, employment, productivity, and labor cost in terms of annual average percentage—change in each of the fifteen European Union countries for: 1974-1985; 1985-1990; 1990-1996; 1990-1994; 1994-1995; and 1995-1996

This report offers a mass of information on the evolution of the economy of the European Union, the problems, and the measures proposed to increase employment in Europe.

2. The European Union has also made an important contribution to the welfare of workers and improvement of labor-management relations.

   In 1996, the European Trade Union Institute published the *European Trade Union Yearbook: 1995;* edited by Emilio Sabaglio and Reiner Hoffman. In fifteen articles, the evolution of the trade unions and their impact has been explained in subjects such as:
   - Trade union regulation in Europe
   - Growth and employment
   - Gender and the labor market
   - European work councils
   - Current trends in collective bargaining in Europe
   - First round of European negotiations
   - The increasing importance of the Interregional Trade Union Councils
   - Social policies and social dialogue in Europe
   - Social protection policies in Europe

- The contribution of community law to harmonization of the working environment
- The European Court of Justice
- Trade union education in Europe
- Improvement of living and working conditions in Europe
- Vocational training in Europe
- The Economic and Social Committee and social integration

The range of subjects covers a wide variety of detail useful in judging the social climate in the region.

THE POLITICAL SITUATION

Or should we call it *political stability?* Times have changed in Europe in the decades since the European Union has taken over some of the national powers. If things do not work smoothly, those in the various European Union countries can blame it on "Brussels" or on "Europe," meaning the European Commission.

Sociology and Human Relations—Story #1

When deciding on a location in a given area, the selection of a general manager could be one of the most important factors regarding the success or failure of the plant. Here is an example set in a European country.

We located a plant for an American multinational in an area between Paris and Lyon, pointing out that one of the problems to be solved was the absence of management skills in that area. Further, the man who was in charge of plant personnel function viewed the extremely left-leaning union as being very hawkish, i.e., very aggressive. The European management decided to appoint a "local boy," an engineer born and raised in that area, as the plant manager.

This new venture was a flop, because the people had no respect for the new man—the "local boy"—and went so far as to gossip about the attitudes of that man's family during World War II.

Finally, a new engineer from Paris was called in to save the venture and peace was soon restored. The new man soon gained the respect of the employees in that plant. The Communist union folded in the face of the strong, authoritative attitude of the new man.

In the three decades after World War II, one studied political stability very carefully in terms of the number of government changes, the composition of the legislature—especially the number of Communists elected, the number of devaluations or revaluations (rare), and inflation and deflation (very exceptional).

One speculated as to what would happen when Franco might die in Spain, what would happen after the Portuguese revolution, and what the effect of neocolonization would be in Western European countries.

It was not an easy task; for example, Italy with the largest number of government changes remained a democracy in spite of its large Communist membership. In fact, in Italy we found that it was easier to achieve results in terms of permits issued and incentives obtained during the period between governments than when the coalition governments were in office.

Sociology and Human Relations—Story #2

A factory was located near the Dutch-Belgian border. It was successful, except for the choice of the general manager.

In the initial phase, an American manager was sent from the midwestern United States to supervise the construction and operational start-up. After a few years, since his wife had never adapted to living outside her country, he returned to the United States. She had made no contact with the locals, nor had she attempted to learn the language or culture.

The successor to the American was a flop because the young local man who became the general manager was not respected. Almost 100 years earlier, there had been a killing incident at the border of the two countries. The victim's relatives carried hatred for all those years against the young man's family. Finally, a new manager was sent from the United States and tranquility was restored.

Portugal quickly turned into a model democracy after the death of Dictator Antonio Salazar and the Anjer revolution. Germany remained stable in spite of the reunification of the two Germanys. France survived the changes in republics and continued to play an important political role in Europe. Scandinavia remained an example of stable government and rational policies, in spite of the high-welfare party policies of the Social Democrats. The population of The Netherlands grew in size and with a quality of life that resulted from a sweeping range of policies that was generally understood elsewhere. The United Kingdom swung from a socialist era to the Thatcher era then back to the Blair social democracy, all without hampering the capitalist growth of its companies. Belgium continues its war of words between the Flemish and Walloon parts of its country, while demonstrating a remarkable stability of its economy that was accomplished despite a huge but manageable debt.

The European Commission, and now the European Parliament, have changed all that. NATO has added to the feeling of security, especially since the changes in the Russian political scene.

The regional differences and cultural mosaic remain. The languages are still different, but the borders are gone. People, goods, and money move freely among the fifteen nations that recently expanded to twenty-five nations. The euro, already accepted by twelve of the fifteen nations of the European Union, is positively affecting the economy and is becoming an equal partner to the American dollar on the international financial scene.

Sociology and Human Relations—Story #3

While we were on our annual sales trip from Belgium to the United States, we had lunch with a CEO in Los Angeles. He complained about the tension between workers and management at one of their European plants. There seemed to be no explanation for the high turnover and absenteeism among the 600 workers at that factory. He asked us to investigate this situation, and we agreed.

In a situation where worker-management tension exists, you can hardly stroll right into the factory and start interviewing. To get insights into the causes of tension, we started by observing the flow of the workers at 5:30 p.m. when the day shift left work. Most of those leaving the plant left on bicycles. We noticed that a group stopped at a nearby café to have a few beers before proceeding home.

To avoid looking like outsiders, we had dressed casually. We went to the café and sat at the counter where we could have access to the workers. We were able to engage in conversations with the workers which we did for three days in a row. We knew the general manager to be a capable young engineer of German descent who was born in the United States. He was sent to manage the plant because of his intercultural skills and knowledge of European languages. Though born in the United States, he spoke those foreign languages with a typical German accent. He was a nice, friendly, young man.

The plant workers, however, had a different perspective of the man. Less than half an hour from the plant, there was a town where an atrocity had been committed toward the end of World War II. As the SS part of the Nazis withdrew from that area, they burned alive some 300 women, children, and elderly people. This action was taken in revenge for the killing of an SS officer by the resistance movement. It is not surprising that the local people hated anything associated with the Germans.

Based on our advice, the young engineer was promoted to a significant role in a new plant in the United States. He was succeeded by his assistant, who was an equally capable young engineer from the local area. Turnover and absenteeism improved immediately. Until now, no one has been told this story about our role in these changes, which resulted in a major improvement in human relations at that factory.

THE ECONOMIC SITUATION

The same remarks made in regard to the political situation are applicable to the economic situation. The same worldly political changes and new generally favorable conditions are leading to economic and monetary stability. Now that free movement of persons, goods, and money is achieved, the influence of the European Union is growing daily. Exchange rates have been established and tied to the euro. One currency is already being used in nearly all the countries of the European Union. American investors understood the value of a single currency sooner than their European counterparts.

Political Interference—Story #1

Political interference with location decisions is widespread. Most companies will not make a location decision when political pressure is likely to be realized. We are not referring to attracting businesses by offering maximum incentives. Rather, we are speaking of measures used to persuade companies to invest. Such measures may include using diplomats, holding back on permits, providing manipulated statistics or incorrect facts and figures, and proposing favors not available to competing companies.

A client was about to make a decision on a site when a female staff member of a government organization arrived for a meeting with the CEO. She brought with her a report that included figures her organization had manipulated against the proposed site. The government report proposed an alternative site in an area that was troubled by social unrest and unemployment. We warned the company that the site suggested by the government was located in an area where the top politician belonged to a party no longer in power. However, the politician had signed the agreement to obtain the permits and incentives negotiated before he lost the election.

The CEO declined to accept the government report, relying instead on information provided by its own consultant. The government ultimately raised so many problems that the plant was built in another country.

Political Interference—Story #2

Not all political interference stories end with the company choosing another country for its investment. If local support is strong enough, there can be a positive outcome.

In one such instance, a site selection led to a comparative analysis between sites and regions on both sides of the border of two neighboring countries. In one country (Country A), a 21 percent cash grant was promised to the company. However, since operating costs were higher in that country, we proposed the site on the other side of the border (Country B) on the condition that a reasonable cash grant could be attained. That area, though, did not qualify (officially) since the local politicians belonged to the opposition party.

We negotiated with national authorities in Country B without success. Finally, we announced to the local press that the new investment would be done across the border in Country A. That announcement caused a storm of protest, because of high unemployment in Country B.

The national administration of Country B went behind our backs to the client in the United States and promised a 17.5 percent cash grant (lower than the cash grant promised by Country A), which is what we had proposed. That proposal sealed the deal, for a happy end to the story for both Country B and the United States company. For us, "all's well that ends well."

Political Interference—Story #3

We once worked with a particular Japanese company that was very successful in selling videos in Europe. The government of one European country learned that this company was looking for a plant site. The country proposed that the company choose a site within its borders. During its site investigation, the Japanese company discovered a lack of customs officers in that country, thus it had a concern about whether delays would occur in getting goods through customs.

The Japanese company exported 10,000 units through one of the cities where there was a lack of customs officials. The customs officers in the European city checked every one of the 10,000 units to see if they were in accordance with the import papers. As a result of this detailed inspection, only ten to twenty boxes were cleared each day. It would have taken almost 1,000 days, or three years, for the entire lot to clear customs.

The Japanese company gave in to government demands to fund sufficient customs officials so that all their units could be cleared within one month. After dealing with this expense, the company dragged its feet on making the decision to invest in the European country. It finally abandoned the idea of an assembly operation in Europe. The decision was made easier because of declining sales and competition from several European companies.

Political Interference—Story #4

The plant in story #2 was built and followed several years later by a second project, which was handled in the same manner as the first investment. The scenario of the company officials' visit was the same as for the first investment.

When our prospective client visited the area (the ambassador in the meantime had been moved to another post), the Japanese managers were still happy. They referred to several problems, the foremost being the lack of specialized labor and experienced management. Labor pirating (one company hiring away employees from another factory) was common in that area, precipitated by the fact that there were now three substantial Japanese factories in that area.

Because of the problems encountered by these earlier projects, we were asked to work for the client as a new and independent advisor. We proposed sites in other countries where those problems did not exist, and successful investments were made there. Our client was very pleased with our approach and the success of the new ventures. That satisfied client became the best source of publicity for our Tokyo office.

Chapter 7

Other Elements

From reading the first six chapters, it must be clear that *location* is a very, very complex science, involving not only economics, but also sociology, psychology, geography, statistics, management, politics, and financial sciences.

FINANCING "THE INVESTMENTS"

There are two ways to finance investments: by using one's own money (OM) or by using the money of other people (OPM).

1. As mentioned earlier, an investor wishes to use the least possible amount of his or her own money and reduce his or her risks, thus he or she tends to look for money from other sources. What is the ratio between OM and OPM? It is hard to say, but as a rule of thumb it is 40 percent (OM) and 60 percent (OPM) in Europe. In Japan that ratio is 10 percent (OM) and 90 percent (OPM). The ratio affects the return on the OM money. Availability and the "cost" of money determine the kind of money used.

    At the end of World War II, cash was scarce and the exchange rate varied considerably. Central banks borrowed from the International Bank and local banks relied on central banks. This picture has changed significantly over the years since the end of World War II. With the creation of the European Union, the European Monetary Unit (EMU) and the euro (European common currency), and the establishment of the Central European Bank in Frankfurt, Germany, the flow of money across borders has been freed. This is not the case when dealing with non-European Union countries in Europe or countries outside of Europe.

The flow of Direct Foreign Investment (DFI) into Europe from the United States and Japan (and the reverse) and investments in countries outside of Europe that are growing economically all result in the creation of a major source of OM. That OM is then used by companies that are expanding. These factors then contribute to the globalization of manufacturing and sales.

2. OPM can derive from many sources.
- Bank loans

   Short- and long-term money is available from banks in all countries. Interest rates vary considerably, depending on several factors. Banks require triple security and provide the most expensive of all sources of funds available. American companies investing in Europe for the first time will rely on the European correspondent bank in their home country. Once the American companies establish subsidiary companies in the foreign countries, however, they learn quickly and are soon ready to borrow money from local banks. They will tend to limit borrowing, however, so they can maintain a healthy ratio of borrowings to OM.

   Japanese companies, on the contrary, are quite accustomed to borrowing from banks. In Japan the ratio of OM to OPM is very highly in favor of OPM. In many cases, only 10 percent of the investment and shareholders' capital is own money. The banks are, as a result, historically large owners of industrial and trading companies. Banking companies are not numerous. Mergers and acquisitions are very common.

- Stock exchange money

   Money to finance investments can also be raised from the stock market. In the growing economies of the United States, Japan, and Europe, the stock exchanges play a major role in providing investment and operating funds to industrial concerns. Institutional investors (especially in the United States) can influence the value of shares being traded.

   Over-the-counter, and more risk-taking dealings in high-tech shares are handled by special exchanges such as NASDAQ. Industrial investors "play" the stock market more and more, especially during periods of economic stabilization or growth.

Various types of bonds, especially government bonds, are less risky forms of investing for small investors. Larger corporations use this method of financing considerably less.
- Subsidies

    Subsidies are one type of incentive that come in the form of cash grants as a percentage of investment in land, building, and equipment (fixed assets). Grants are nonrepayable and, in some countries, nontaxable.

    Grants can be nontaxable and nondepreciable; nontaxable and depreciable; or taxable and depreciable. These differences can have a significant effect on the return on money invested.

    Subsidies can be given in the form of reduced interest rates on invested capital. The additional employment resulting from the investment will, in general, affect the size of the grant and the size of the reduction in the interest rate. Labor-intensive investments benefit significantly from this form of financing.

Generally, the three methods of financing are used in combination, not in one form only.

CASH FLOW TABLES

1. The purpose of using cash flow tables in making the location decision is twofold.
    - To calculate the magnitude of the differentials between various cost factors (nonrecurring and recurring cost factors) applicable in the *preferred areas* or projected on the *preferred sites*.
    - To compare the rate of return on investment of OM, the profitability index, and the payback period of the investment, projected on a number of preferred sites.

    The calculations are intended to provide an order of magnitude, not a clear-cut representation, as figures change according to evolving requirements and time of calculation. No discounting to allow for inflation is provided in these tables.
2. Table 7.1 and Table 7.2 illustrate the use of cash flow tables for various investment projects.

3. The relevance of the cash flow tables is best demonstrated in feasibility studies in which projections of the financial aspect of investment projects in various locations are made. These projections are made in order to reduce the number of preferred regions or cities.

INDUSTRIAL LOCATION AND ECOLOGY

Environmental policies have an exceptional impact on location investment decisions. Indeed, ecology is a location factor that is (1) a *nonrecurring cost* because of the equipment investment required to clean the air and/or water; (2) a *recurring cost* because of the *operating cost* incurred when using products, power, personnel, etc.; and (3) an *intangible factor* because of the impact of legislative require-

TABLE 7.1. Suppliers of Direct Investment of the European Union—Inflows

Year	United States in European Union	Japan in European Union	EFTA[a] in European Union	Total European Union Inflows
	(million euro)			
1987	2.237	1.572		12.991
1988	2.551	2.584		18.141
1989	9.846	4.354		27.943
1990	9.178	5.406		32.753
1991	5.411	1.682		20.933
1992	12.286	1.859	3.303	22.760
1993	11.296	1.600	2.902	21.504
1994	10.347	1.454	5.630	21.814
1995	24.293	1.535	7.064	37.220
1996	15.931	958	6.285	28.420
1997	22.401	995	937	35.970

Source: Eurostat: Panorama of the European Union, 1987-1997.

[a]European Free Trade Area (EFTA)

TABLE 7.2. Suppliers of Direct Investment to the European Union—Outflows

Year	United States in European Union	Japan in European Union	EFTA in European Union	Total European Union Outflows
		(million euro)		
1987	23.885	−12		30.670
1988	22.120	247		31.680
1989	24.053	682		33.282
1990	7.155	911		20.527
1991	9.232	341		26.732
1992	6.941	445	1.539	17.828
1993	13.789	−1.229	1.758	24.157
1994	7.426	272	4.279	24.129
1995	24.534	854	1.797	45.580
1996	13.207	1.822	4.794	42.766
1997	33.542	248	6.376	77.671

Source: Eurostat: Panorama of the European Union, 1987-1997.

ments and the application thereof—on the two categories of costs mentioned previously.

In the decennia after World War II and the liberation of the Eastern European countries from the communist regime, ecology played a very small role in determining the location for an investment. Ecology plays a small role still today in countries in the developing world. While ecology plays a small role in developing countries, it can play a decisive role in developed countries.

In recent decades, due to the appearance of Green parties in the political scene, ecology has become a very important factor in deciding the location of industrial manufacturing projects.

IN CONCLUSION

The three location decision-related elements, or factors, dealt with in this chapter serve only to demonstrate what one faces in looking for a suitable—or best—site. Local people can reveal the existence of undetected or undetectable pitfalls.

A covered dump, smoking lightly in the summer heat, will not be a good site for a pharmaceutical plant. Believe it or not, a consultant recommended just such a site.

Chapter 8

Community and Site Analysis

The economic analysis of a proposed investment has been outlined. The next step is to analyze the region, the geographic area that is defined as the *preferred one* in the area of search specified in Chapter 3. Before establishing the physical fit, one must analyze the region in order to identify the factors that may have an effect on the location decision. (See the Community Analysis worksheet at the end of this chapter.)

GEOGRAPHIC ASPECTS

The physical and economic geography aspects are part of the regional analysis. Siting an economic activity means to operate under conditions that are in many ways determined, retained, or assisted by the location in terms of geography. The following are some of the key components of geography.

1. Means of transportation, such as roads, highways, canals, railroads, airports, and ports have an impact on making decisions on locating investment projects.
2. Infrastructure—especially physical infrastructure—distances, population concentration, industrial concentration, etc., provide assistance to economic activities.
3. Soil conditions and ecology also need to be studied in the regions to determine their effect on costs.
4. Mapping and calculating distances, density of population, demography, and labor mobility must be considered and evaluated, particularly when dealing with large labor projects or large employers.

> **Importance of Infrastructure—Telephone Lines**
>
> In the late 1960s, we were involved in a plant location study along the Mediterranean Sea. One particularly important factor in our analysis was the phone infrastructure. We needed to understand the availability of telephone lines, speed of installation, and costs of equipment and phone calls.
>
> We went to the regional phone company office in a major port city. While waiting to see the supervisor, we met a Japanese fellow who was sweating over a bunch of papers that needed to be filled out. He was a representative of Mitsui, a trading company that wanted to set up a logistics center with a battery of telex machines. This center would handle the cargos of the Atlantic Ocean and Mediterranean Sea. Since his French was insufficient to answer the questions on the form that needed to be completed, I volunteered to help him.
>
> This representative needed twenty-nine telex and telephone lines within sixty days. Once we completed the forms, the man handed the papers to the clerk who disappeared for about ten minutes. On his return, he said, "I checked with the planning department. You may come back in seven years."
>
> The Japanese man looked puzzled until I explained that there was an implemented five-year plan for the overall expansion of the number of telephone and telex lines in the region.

SOCIOECONOMIC ASPECTS

Both sociology and economy influence the results of projected investments in a significant way.

1. Attitude and behavior are sociological aspects that affect buying and being employed in a region or community.

In the mosaic of regions in Europe, types of human behavior vary from one region to another. As a child, we remember how our mother talked about the town in which our father was raised, as well as the town where she was raised. For decision making, it is important to understand these local perspectives.

One story concerning local sociology takes place in Central France. One of our American clients located a new plant in a town near the Loire River. A French engineer who was originally from that town was selected as the plant manager. The man had absolutely no authority from the get-go. Everyone referred to him by his first name and joked about his humble origin. He was the son of "so-and-so" who no one liked or respected.

His local origins worked against him in that position. We recommended a change in management, and appointed an authoritarian French engineer from Paris. Once the new engineer was introduced to the operation, the plant became very manageable. Many such stories can be told relevant to practically every country or region we studied.

2. Unions

Several American companies asked us to look for sites in regions in Europe where there were no unions. We told them that this was impossible. It is necessary to learn to live with unions, to know "who is who" in the labor movement, to understand why people join or live with unions, to know which unions were revolutionary or militant, which were cooperative, and which were attached to political parties or religious groups.

Lack of understanding of union-management relations has caused considerable hardship to U.S. investments in Europe, especially in the 1960s and 1970s. Further, union relations generally are defined in labor relations laws that vary from country to country. Was it not Margaret Thatcher who would break the necks of unions in the UK?[1]

3. Unemployment

The economic analysis of a region is generally reflected in the unemployment figures. The degree of employment is the measure of success for a pension fund (high employment means less payout from the pension funds). The rate of unemployment is the "temperature" of the people's willingness to work. When unemployment compensation is close to the minimum wage scale, the desire to work is low, and the number of black market workers is high. With a high social security level of payment covering unemployment compensation, the degree of willingness to work is even lower.

---

[1] Margaret Thatcher, then Prime Minister in the United Kingdom, limited the power of labor unions by passing restrictive laws.

4. The Community Analysis

The standardized community analysis shown at the end of the chapter provides a picture of the climate in which a new investment will operate. Matching between the *offer* of the community and the *demand* by the investor determines the likelihood of a successful plant operation. Strained relations between the community and the company will ultimately lead to failure of an operation. On the other hand, a high degree of cooperation will contribute in a significant way to the success of a plant investment.

SITE LOCATION PROCESS

Is there truth to the saying "all prairies look alike"? Is there no such thing as an *ideal* site? Yes—there is no *ideal* person, nor an *ideal* situation, nor an *ideal* community, nor an *ideal* site. All sites have advantages and disadvantages. The selection of a site is always the result of some compromise of factors.

Many location specialists start by looking for "a suitable site." We differ because looking for a site is the last—not the first—thing we do. First, we look for the "best" or most suitable region—the *one that comes the closest to meeting the company's requirements.* We look for the most suitable town in the region. Only then do we look for the most suitable site. If that site is not available, we endeavor to *make it available.*

In recent decades, we have applied this procedure many times. If a community is convinced that the project is an asset to its people and if one specific piece of land is most adaptable to the needs of the company, that community will be able to convince the company owners to select its site and use compulsory purchase powers to make it available.

The site analysis sheet at the end of this chapter shows the physical and economical aspects of a site that should be considered. A site is *more than a piece of real estate.* It should also be noted that environmentalists are imposing more and more restrictions on land use lately.

Warning: Watch for the matchmakers in some countries. Remember the sad stories in Spain relating to matchmakers? Here is another story about matchmaking in Belgium.

Matchmaking in Brussels

An American company wanted to set up a large distribution center near Brussels. When we met with the company's management at its headquarters in the Midwest, we were astonished to hear of the high cost for the land being quoted. When the treasurer wound up getting a receipt for only half of the price paid for the land, we became suspicious and investigated the situation.

It turned out that the local manager, who knew the middleman and was a friend of the regulatory body of the town, shared half the price in the form of an advanced payment, without a receipt of course! Someone lost his job that day, and the story has made the prospective buyers more vigilant.

COMMUNITY ANALYSIS

Location

1. Character of town: industrial, commercial, touristy, etc.
2. Historical background (origin, economics)
3. Exact location with reference to major cities (North, South, East, West)
4. Major distances to major cities

Transportation

1. Roads
   - Existing and projected roads and programs
   - Highways and major roads (state and regional)
   - Limitations in height or weight (e.g., bridges)
2. Railroad
   - Nearest station for passengers or goods
   - Number of destinations and connections
   - National and international railroad lines
   - Type of freight station, kind of handling equipment, and maximum loads
   - Railroad siding: existing, possible, installation cost, and timing

3. Waterways
    - River or canals
    - Maximum load of barges
    - Width, depth, height of bridges
    - Flow
    - Can the water be used for cooling purposes and fire protection?
    - Is the canal fed by pumps or a natural supply of water?
    - Is there a dock or is it possible to have a dock along the waterway, and what are the specifications?
4. Seaport
    - Activity: general cargo, container, roll on/roll off, etc.
    - Maximum size of ships
    - Frequency, type, and destination of regular lines
5. River port
    - Activity: general cargo, container, etc.
    - Equipment
6. Airport
    - National or international
    - Yearly traffic of passengers and freight
    - Frequency and destinations of major flights
    - Flight schedules
7. Population, industrialization, and employment should be dealt with separately for the following areas:
    - The city, town, or village
    - Area within a radius of 5 to 10 kilometers or 15 to 20 kilometers
    - A well-described labor drawing area

Population

1. Current number of inhabitants (male, female)
2. Evolution and projection of population
3. Population growth (birth rate versus death rate)
4. Emigration and immigration
5. Population age structure

Industrialization

1. History of industrialization
2. Existing industry: names, products, and number of employees for the major companies employing 100 or more

3. Evolution of the industrial activities (expansion or decline)
4. New planned industries, estimated employment, type of activity, etc.
5. Name, activity, and employment of all industries located on the proposed industrial site
6. Type and name of industries that stopped activities recently or intend to stop, and the number and qualifications of those freed

Employment

1. Active population and evolution
2. Rate of employment in the following three sectors:
   - Agricultural
   - Industrial
   - Service
3. Availability of labor (female and male) due to:
   - Unemployment
   - Immigration
   - Balance of outgoing and incoming commuters
   - Decline in industry
   - Labor freed from agricultural activities
4. Type and qualification of available labor (The description of the available manpower should be clear to have a good idea of the number of employees who are unskilled, semi-skilled, and highly skilled.)

Community Facilities

1. Housing
   - Availability of houses for middle management
   - Rentals
   - Building cost
2. Education
   - Primary and secondary education
   - Technical schools, high schools, colleges, university facilities
   - Number of pupils, students, and annual school dropouts (For small communities or cities, please indicate the kind of education available in the vicinity.)
3. Cultural and recreational facilities
   - Cultural: theaters, museums, libraries, conference rooms, historical attractions, etc.

- Recreational: sports activities (indoor and outdoor), entertainment, recreational parks or natural parks in the surroundings, tourist attractions, etc.
4. Hotels and banking
    - Hotels: number and type of hotels, number of bedrooms
    - Banking: main offices or subsidiaries, national or international (names of international representation)
5. Medical facilities
    - Description of the health care in the community and surroundings or in the nearest big city
6. Others
    - Is there a Chamber of Commerce?
    - Are there conference rooms for meetings?
    - Is there a Lions or Rotary Club?

Industrial Site

The site description should contain the following:

1. A location map (i.e., a detailed road map with indication of the site)
2. A site map showing the exact limits and/or expansion possibilities of the site, with indication of the site's existing features
3. A site analysis, such as the one following this questionnaire, examining plants and available land

SITE ANALYSIS

Site Description—Identification

Location

1. Country
2. Province/state/department
3. Arrondissement/county/kreis
4. Town/city

Physical characteristics

1. Area (in acres or hectares)
2. Shape

3. Topography
   4. Elevation (difference in)
   5. Drainage
   6. Flood problems

Soil conditions

   1. Type of soil
   2. Bearing capacity
   3. Underlying soil
   4. Undermining
   5. Ground water level

Environment

   1. Present use
   2. Zoning and planning (light, mixed, or heavy industry)
   3. Boundaries
      - North
      - West
      - South
      - East
   4. Expansion potential
   5. Neighboring industries
      - Name
      - Activity
      - Employment
   6. Vicinity remarks
   7. Prevailing winds

Site preparation

   1. Clearing
   2. Removal of structures
   3. Grading

Site cost

   1. Per square meter (m^2)
   2. Per acre
   3. Total site cost
   4. Ownership

INFRASTRUCTURE

Transportation

Roads (existing or projected timing and eventual traffic problems)

1. To the proposed site
2. On the industrial estate
3. Distance to:
   - Highways (also, access of)
   - Town
   - Freight station
   - Airport
   - Harbor

Railroad

1. Position of line
2. Siding
3. Nearest freight station

Waterways

1. Type
2. Distance to site

Pipelines

Telephone and Fax Availability and Reliability

Internet Capacity Availability and Reliability

Utilities

Power

1. Power company
2. Position of line
3. Nearest substation
4. Incoming voltage
5. Air, ground, closed circuit

Gas

1. Gas company
2. Type of gas (Kcal)
3. Location of the nearest line

Water

1. Sanitary water
   - Water company
   - Location of the main line
   - Pressure
   - Capacity
   - Analysis
2. Cooling and process water
   - Source
   - Capacity
   - Distance
   - Analysis

Sewerage

1. City sanitation
   - Location
   - Capacity
   - Special requirements
2. Cooling and process water
   - Location
   - Capacity
   - Special requirements
3. Storm sewerage
   - Desired
   - Unnecessary
4. Treatment plant
   - Location
   - Existing or projected
   - Total capacity and availability
   - Type of treatment plant

Hiding the Deficiencies of a Site

When assessing sites in the Pisa-Livorno area of Italy for a mechanical plant, we determined that infrastructure and traffic logistics were very high on the list of requirements.

The local authorities, including the public works department, provided full cooperation. Several good sites were identified and analyzed. Many more, some of which were identified by the city, were considered and rejected. During the last checkup, the public works manager suggested a site that was located strategically near two projected highways. The selection of that site would provide an excellent opportunity to publicize the new plant.

Three sites, including the one suggested by the public works manager, were being considered. We told the official that our client could accept none of the three sites until boring and bearing capacity studies were conducted. He assured us that this would not be a problem, since he had already performed those studies. He provided our client with the results.

We visited the proposed site around noon the following day. It was steaming hot. We walked over the site and noticed an unusual acidic smell, and some fumes here and there rising up from the subsoil. We learned that the site was previously a city dump, which was why the city was willing to sell it for less money.

That site was discarded from consideration.

Chapter 9

Decision Making

Charts and figures needed for locational decisions are not just a bunch of vague or useless statistics. The previous chapters have tended to focus on information needed to make a decision. At the end of the *analysis,* we must arrive at a *synthesis.* This is the most difficult and least understood part of location studies, and requires all the experience one can muster.

THE WEIGHING OF FACTORS IN LOCATION DECISION MAKING

There are various approaches for weighing the factors involved in the location decision.

1. The most common approach is preparing an evaluation using a one to ten scale. It is the most subjective method of all and can be used in the negative or positive sense. Remember, "the glass can be half full or half empty." Furthermore, company policies, dominating characters or biased nationals will certainly "take a ride" with this method. Have you not noticed that the national or ethnic origin of the investigator dominates the judgment? A French manager of an American company that intends to locate a plant in Europe will certainly *weigh the factors toward France.*
2. Weighing the factors will involve a step-by-step elimination process. If you have not kept an open mind and looked at all the probabilities, you will certainly have lost a valuable part of the arguments in favor of or against certain areas.

Locational Site Selection Bias

Once, on a plane traveling from the United States to Europe by way of London, we sat next to a family. The father's job was to look for a plant site in Europe to start up his company's first production unit outside the United States. He and his wife were of Irish descent, and their children certainly didn't look any different from their Irish parents. The family was on its way to Dublin.

I asked them why they were choosing Ireland. The answer was clear: they were Irish, had lived in Ireland earlier for one year, had never traveled to the continent and could speak only two languages—English and Irish. These were all the reasons involved in their choice of Ireland.

Postscript: The plant was built in Ireland but closed three years later. Ireland was the wrong place for that specific project.

A plant is not built to last forever, but an investment—once made—cannot be moved when a decision is found to be erroneous.

The most difficult aspect of weighing the factors is judging the many rating systems that tend to evaluate investment climate. Trends in making locational decisions will tend to be more or less colored by the training of the expert evaluator.

Japanese Decision Making—Story #1

For a long time, the Japanese clients we worked with were not very oriented toward using consultants. Usually it took a couple years of experience and some horror stories before they would consider using a consultant on a project. Even then, they almost never provided the full basic data needed for a consultant to make a sound analysis and evaluation.

The concern these men expressed was their fear of espionage. They feared that a consultant would pass on important shared information to competing companies in Europe and America. On the other hand, when these same men saw the names of competitors on our client's list, they tried to obtain information on machinery, figures, and confidential items of those companies. We had to remind them that we were in the location business, not in the business of espionage or public relations.

The same attitude toward outside competitors existed among the Japanese within various departments. One morning, we presented a location study report to Division A of Company C. During lunch, we were requested not to tell their colleagues in Division B (with whom we were negotiating a study contract) that we had worked with Division A.

In fact, the results of our study were never shared between divisions A and B. Years later, I understood the reasons for this attitude after the events of a study described in "Japanese Decision Making—Story #2."

DRAWING CONCLUSIONS AND MAKING RECOMMENDATIONS

Drawing Conclusions

The evaluation of the location factors and the weighing thereof as a means of establishing the order of importance must lead to clear and unequivocal conclusions. The conclusion should never point to one site, in one region, in one country. Remember again: there is never only *one solution* in selecting the optimal site. *It is always a compromise of factors.*

Japanese Decision Making—Story #2

Many years ago, we were asked to locate a billion-dollar Japanese plant in Scotland. During our presentation in Tokyo, we noticed that the team of six attendees was taking many notes and then comparing the various pages with a four-page memo in Japanese.

During an intermission I asked our office manager what was happening. He said that the Japanese team had its own questionnaire with a large number of important items that were to be considered for the location decision. The office manager obtained a copy of that text and mailed me an English translation of it. Immediately, we called the town mayor in Scotland to report on our presentation in Tokyo. He indicated his hope that the Japanese company would decide in favor of his town. We also mentioned the existence of the Japanese questionnaire and proposed that we should prepare the answers to that questionnaire in a little separate brochure before their visit to Scotland was announced.

Three or four months later I received a request from the Japanese to meet them in Scotland, so that we might visit the proposed site together. I met their flight on a Monday morning and drove them in a little rented bus to the town's city hall. The mayor received us and welcomed the Japanese group. He expressed the hope that they would find happiness and profitability in his town. The eldest of the men thanked the mayor and opened the session for questions.

When we saw the team take their questionnaire out of their bags, I suggested to the mayor that we distribute the summaries we had prepared in advance. The Japanese looked at these summaries with surprise and a concert of "ho-ho-ho," the Japanese expression of approval, followed. After going through the brochure for fifteen minutes, they said, "No more questions," then turned to us and asked, "How did you get those questions anyhow?" A smile was our answer.

The remarkable thing about the questions was that they expressed the typical concerns of a Japanese company specific to its part of the world, such as:

When was the last earthquake?
How much damage was there from the earthquake?
How many victims?
Do we need special insurance?
Are there still active volcanoes nearby?

We can laugh, since these questions were posed in Scotland, but to the Japanese these were serious issues. About noon, the team was ready for the planned site visit. However, since we were all thirsty, we stopped first at a very, very old pub. The men enjoyed that pub so much that they consumed several beers before we moved to lunch at a nearby (also very old) typical Scottish restaurant.

It was well into the afternoon before we visited the area and the site in our little bus. At the site, the Japanese brought out their cameras (each visitor had at least two cameras) and took many photos. We still cannot understand how it is possible to take that many pictures of a fifteen- to twenty-acre prairie lot. At the end of the day, we discussed our plans for the remainder of the week. The Japanese could not return to Japan before Saturday, since they were given one week to complete their assignment. I then suggested the following: either I transport them to one of the many golf courses nearby, or they could join me on the eight o'clock flight to London in order to go shopping. The decision was unanimous—go shopping. So we all left on that flight to London, then I continued on to Brussels.

For three months we heard nothing despite my frequent inquiries to our office in Tokyo and the many impatient calls from the mayor of the city in Scotland. Then one day we received a telex, which directed us to meet the new team next Monday morning at the airport. It happened in the same manner and was an exact repetition of the first team's visit. You may not believe this, but a third team, this time made up of seven members, arrived a few months later. The third team announced that the first two teams could not make a unanimous decision and therefore it was necessary to send a third team to Scotland. In each case, the visits followed the same pattern.

Shortly thereafter, we were invited to Tokyo for a final roundup, since the team had come to a favorable conclusion. The top bosses of the Japanese company wanted to ask some more questions about the project. During the dinner at the geisha house, we asked our long-refrained question: "Why not the first and second teams? Why a third teram? Why six people in Teams One and Two, but seven in Team Three?" Can't you guess?

The team must decide without reservations. If anything should go wrong at a later date, it would mean the end of a career for the company. Each team member had a specific task and specialty—a personnel man, an engineer, a finance man, a lawyer, a marketing man, and a secretary.

The seventh man on the third team was the son of the president of the company, and his role was to make sure the team would arrive at a decision without fear of failure.

Make it clear that if the company wants to achieve one goal, several possibilities exist to reach that goal. In an important study for a major American automotive company in France, conclusions led to the following reasoning: In view of the large site area needed for a plant that would house 10,000 workers and serve the Western European market, the right skills needed to be found in the areas selected for study. The location study could take several directions.

1. Assembling, subassembling, and the manufacturing of mechanical parts could be located on one large 1,000-acre site. Only one area in France could be found, however, where sufficient labor was available. Unfortunately, that area was known for its militant labor union that caused the closure of what had been the main industry in the area.
2. Dividing the production functions described previously would have been more costly, but could have led to fewer labor problems. With this alternative, it would have been possible to assemble and subassemble in one area, while the mechanical parts

could be manufactured in another region with skilled labor better adapted for that purpose.
3. Since the market required flexibility in terms of the geographic area served, it would be possible to centralize the assembly and subassembly work near a major Mediterranean port where steel sheets were manufactured and the mechanical parts inland (less than 100 miles away). Alternatively, a site in eastern France was available for assembly and subassembly. The mechanical parts could still come from the region previously mentioned.

Under pressure from the influential French automotive industry, the French administration opposed the plan. The motors and other mechanical parts were produced in Germany and Austria, while the assembly and sub-assembly plant was set up in Spain.

Fortunately, much has changed in France since the European Union blurred nationalist feelings. Also, authorities are now aware that competition is a positive factor in raising the standard of living for people.

Making Recommendations

The successful implementation of an investment project, after an objective scientific study as outlined in this book, will depend on the clarity and the precision with which recommendations are made to the company facing the location decision.

Outlining the *positive action* for the implementation of the project should be sufficiently complete and logical so that the manager in charge will make no mistakes. The *negative actions* should also be considered. It is as important to enumerate what to do, as it is to warn what not to do.

Describe, step by step:

- How to acquire the site, how much to pay, what taxes are included, what infrastructure is included in the price, and what is *not* included (possibly not even mentioned)
- Building permits, regulations, appointing the architect, the use of a construction engineer
- The permits: operational permits; environmental protection permits on the acquired site and on the operating process
- The financial agreements

- The incentives application
- The hiring and firing procedures
- The lawyer who gathers all the items listed above and puts them into a legal contract

Nothing should be forgotten or omitted. Leave little room for later interpretation so as to avoid confusion and subjectivity.

> **Promises, Promises, Promises**
>
> As previously pointed out, officials of regions and countries will do almost anything to get industrial investments. We have warned time and time again in our recommendations that commitments in writing should be obtained. Authorities in Italy and Spain were among the most difficult to hold to what they had promised.
>
> Among the many stories was the experience of a major automotive assembly plant in Spain. The regional authorities, in order to get the investment on the proposed site, indicated in writing that they would take care of preparing the access roads to the plant and develop a treatment plant to ensure sanitary treatment. The plant was built on time but the authorities dragged their feet for budgetary reasons or because of problems with getting permits from the national administration.
>
> Delays by the regional administration to provide the access roads resulted in severe traffic for the plant's 4,000 personnel. The problem was most critical when the majority of the plant personnel was arriving or leaving the factory. Visitors also experienced great difficulty in getting to the plant.
>
> Because of the warm climate in southern Spain, one could smell the plant even before it came into view because of the frustrating odor of the toilet waste. That waste was piped into the open air for several miles. This problem was a result of the delayed construction of a treatment plant.
>
> It took at least four years before both problems were solved. Even then, the resolution required that the company contribute—in taxes—substantial amounts of money to the regional authorities to help pay for the necessities.

THE USE OF CONSULTANTS

Serving more than fifty years as a consultant and professor, and teaching young people from all over the world about industrial location and regional development have taught me the advantages and pitfalls of using consultants in deciding on the location of investments.

The standard joke goes: "If you don't know enough about the topic, teach it. If you know little or nothing about the subject, become a consultant." In our long careers we have come across examples of both in our field. Why is it that you can probably count the number of location consultants in this world on ten fingers? There are two reasons.

- The complexity of the subject
- The wrong understanding of the field

Deciding upon the location is not an easy task. It requires insight into a lot of aspects that do not all belong to the same discipline: geography, economics, sociology, political sciences, and statistical methods, to name just a few. It is clear that the people who have the following three characteristics will give the advice most likely to lead to success: *experience, experience, and experience.*

To build up experience, one must be consulted and receive the confidence of the company's managers and officers faced with making the location decisions. To be able to offer valuable advice, one must deal with many complex situations.

The "image" of the locational decision process has been, and still is today (after almost half a century later), one of "corporate real estate."

Ever since the creation of the International Development Research Council (IDRC) (formerly called the Industrial Development Research Council) of Atlanta, Georgia, in 1962, the real estate function of the location has been stressed. A variety of backgrounds were, and still are, found among managers dealing with the location problem. These men may be quantity surveyors, architects, engineers, or real estate agents.

Later, the creation of the geo-economics section recognized the value of the site selection process. A variety of procedures, questionnaires, and approaches have been found among facilities managers. All these have their merit but few follow the scientific approach and procedure outlined and described in this book.

And what about the use of consultants? From 1960 through 1989, Plant Location International of Brussels, Belgium, which I founded and managed until it was sold, held a near monopoly in Europe. We consulted American companies primarily, but also European and Japanese businesses. Once in a while, one of the larger American consultants was hired to deal with either locational or regional development projects. Few projects made any money, and eventually use of the service was discontinued.

In the past decade the top five consultant companies have indicated that they have a location division; however, because of the lack of experienced researchers, their operations have fallen into the real estate or general economics trap. The past century has also marked the development of information, communication, and technology with vast data spreads via the Internet (valid or not, subjective or not, useful or not).

With the globalization phenomena, population expansion in Asia and Latin America, and the development of excess production capacity of certain consumer goods (such as cars), it has become necessary to rationalize and concentrate operations. In addition, with the need to maintain competitiveness and make money, there develops an evolution toward megamergers.

Now, as we have entered the twenty-first century, the field of location is changing rapidly. In the past five decades most of the expansion and new product development in Europe was done on the so-called greenfield sites. In the past two decades, mergers and acquisitions have been the order of the day. For a few years, we have a fourth way of expanding and beating the competition—*clustering.*

Consultants must look at the best approach to solve locational problems in light of economic evolution. By "clustering" one refers to a *geographic* and an *economic* feature. In terms of geography, clustering means to concentrate a number of production and/or distribution units in one (generally large) site, for example, to group Tier One and Tier Two suppliers to the automotive original equipment manufacturers (OEM) companies.

In terms of economics, clustering means operating joint subassembly plants (in the case of the automotive industry) whereby a number of component manufacturers will develop new products and manufacture and assemble a subunit that can then be moved (by conveyor belt) onto the assembly line, or shipped to one or more OEMs. The phenomena of clustering can also be applied to other economic sectors such as fine chemicals and some pharmaceuticals.

The new role of the locational consultant is to put a lot of research into clustering development, a practice that is playing a major role in Europe.

IN CONCLUSION

It will be difficult to find the right consultant to assist in the location or relocation studies, yet easy to make a wrong selection.

Choosing a Consultant—Story #1

A one million square foot food plant was to be located in Europe. The international vice president solicited the assistance of a large American construction company to help choose the best site for the plant. In the 1960s and 1970s most of these large companies had in-house location specialists.

By coincidence, we were trying to sell our services to the U.S. headquarters of that same food company when we learned about this project. We had the opportunity to look at the site report that had been prepared by the construction company. The report analyzed only one site in the given country. We pointed out not only the limitations of the study for making that management decision, but also the many serious problems in that area (which we knew about as we had previously conducted a study there).

We decided to make a proposal to the international vp that would have a fee equal to or lower than the fee of the construction company. We suggested that the construction company return the fees it had received from our client, so that company might still be considered to do the actual construction of the plant. The construction work, rather than the consulting, was after all the core of that company's business.

Upon my return to Europe, we met with the general manager of the construction company. That company accepted my proposition and returned the money to our client. We did an excellent job on our study, which was followed by another half-dozen assignments for this client thereafter.

The construction company was rewarded for its honest handling, fired their so-called location man, and received the job of building the plant all to the great satisfaction of the food company.

Choosing a Consultant—Story #2

After two decades of serving many large American companies, we began aiming for the acquisition of a "big name" company that we could use as a reference. During a sales trip to New York, we met the president of a company in the ICT (computer technology) field. He explained that they had a team of twelve people investigating Europe for a large plant that would need a significant number of highly skilled workers.

The report that the previous location group had presented was not entirely satisfactory to management and they were seeking a second opinion. Though our experience impressed the president, he was still hesitant about the potential value of our services. Our offer was as follows: if we were given the same basic information about the project that was used by the company team, we would come up with a new report within eight weeks to facilitate the company's hurry to make a decision. Furthermore, we quoted an all-inclusive fee with the restriction that we were only to receive payment once our recommendations were accepted. Taking business on that basis was a scary experience, and we never repeated that approach in the future.

We kept our share of the bargain and in eight weeks we made a presentation of our study. The president, the international vp, the chief financial officer, and the team of twelve attended that presentation. We had prepared an executive summary and a projection show. After almost two hours of questioning, the president stood up and said, "Gentlemen, I think we have our answer. Professor, come up to my office for lunch and pick up your check."

Choosing a Consultant—Story #3

A relocation study proposal was requested for a medium-sized American affiliate company in Europe. Before making such a proposal, we interviewed the local management on the background of the operation, the company's investment policies in Europe, and its future plans.

We then visited the plant. Unfortunately, access to the plant was terrible. It was located off an inferior narrow road right near the community church. Had there ever been a fire at the plant, the church would have gone up with it. With the presence of inflammable products at the plant, this could have happened at any time. Because of the risk, the insurance cost was very high. In addition, much of the plant machinery, which was of pre–World War II origin, was either not being used or was in need of repair. Because the plant was located in the middle of a small town, it was a checkerboard of small units put together over four and a half floors. The condition of the building and equipment was deficient, as was the condition of the workers.

This plant should have been relocated forty years ago. It's a wonder it was still operational at all. The plant was losing money, of course. The main competitors to the company were located in China.

Our proposal indicated that the company should relocate in the same region (a condition imposed by the union) on a greenfield site at little investment of its own money. The site and the building could be placed in a fifteen-year lease-purchase deal with modern equipment (half a dozen new pieces of equipment would replace the sixty old rusty chaps) and incentives to cover the relocation and moving costs, all to save the remaining fifty workers.

The American manager wanted to keep the same workers because he was afraid of the union's wrath. Besides experiencing several strikes, he twice suffered a forty-eight-hour lockdown in his own office without heat or electricity at the union's hands. He had already sent his family back to the United States.

The unions and the workers voted against the relocation, the manager resigned and returned home, and most of the workers were pensioned off. That is the last we heard about that situation.

Chapter 10

Locating Warehouses, Research and Development Centers, Headquarters, and Call Centers

The emphasis thus far has been placed on the location of production units. However, the rules and procedures set forth in previous chapters applied to all economic activities. The recurring and nonrecurring cost factors are weighted differently in location decision making for each different economic activity.

WAREHOUSES

A warehouse can be part of a production unit and located on the same site. That was the usual case until about ten years ago. In recent years, globalization, megamergers, and acquisitions have changed freight and warehousing to logistics. Two factors dominate when dealing with site location for a warehouse: *transportation cost* and *real estate investments*. Other factors, such as labor and material handling (now mostly robotized and computerized) costs are no longer determining.

Transportation Cost

Inbound and outbound freight are the principal distribution costs and most important for inventory control. The practice of grouping by logistics companies in regional warehousing is increasing. Road handlers are competing for freight and warehousing from the industrial producers as a means to decrease distribution costs.

Road transport, the main means of transportation in Europe, is clogging up the main highways and will cause longer transport times. Continuous increases in fuel cost (mostly taxes rather than basic production and distribution costs) will make water transportation more attractive again in the near future.

If the railroad companies in Europe could be further privatized, integrated, and better organized for freight transportation, trains would be able to move much of the long distance freight economically. Drier port stations (for example, on existing switching yards) are needed, as well as better coordination between the national railroads.

Real Estate Investment

Real estate investment in sites and buildings is a second factor in determining location of a warehouse. Financial formulas and better depreciation policies are able to spread the cost of real estate investments over longer periods, thereby reducing occupancy costs. Corporate real estate facility managers are trying to reduce their companies' investments in bricks and mortar.

RESEARCH AND DEVELOPMENT LABORATORIES AND RESEARCH CENTERS

Two factors determine the location of research and development (R&D) centers.

Headquarters Location and Warehousing for Spare Parts

An American company asked its European manager to look for a site in order to build a headquarters and logistics center for spare parts storage. The requirements pointed toward a major city not too far from a major port.

Everything went smoothly, the ideal site was found, and the request for funding was sent to the United States headquarters of the company. It just so happened that I was meeting with the chief executive officer (CEO) to make a presentation for our plant location services. I was asked the usual questions about our methods, procedures, experience, and references. At one time, the chief financial officer (CFO) asked me if I could explain why some pieces of land in the suburbs of the selected city were unusually costly, and why the European manager was unable to produce a receipt proving that the payment had been made. Some explanation had been given to him that sometimes professional fees and commissions were paid without a receipt. The CFO was unhappy with that explanation and asked me to investigate the situation.

I went to the suburb where the proposed site was located and where the conspiracy was set up. The conspiracy involved the real estate dealer, landowner, architect, European manager of the company, and notary. The conspiracy passed a deed that was only 40 percent of the price paid (supposedly to save taxes); the remaining 60 percent wound up divided among the conspirators. While it was too late to undo the deal, as you might imagine, the European manager was fired, the architect was not appointed again, and the problems were swept under the carpet.

The People

Available educational facilities must be considered first. After all, there would be no Silicon Valley without Stanford University. The innovative ideas and beginnings of many new enterprises over the past several decades came from the brains at universities (professors, teams of faculty and students, and basic research).

University studies are not the end of the education process. More attention should be given to primary and secondary education as well as technical schools. The latter are no longer considered the "waste bin" for brains, i.e., the dumping place for students unable to complete secondary school programs.

Facilities that provide continuing education in and around universities are becoming more valuable since knowledge is no longer static. Knowledge is in a state of continuous evolution, as many professions disappear and new professions are born. The electronic revolution, computer use, the Internet, numerous intranets, the development of the CD-ROM, networking, and the evolution of television, mobile phone, telephone, fax, and e-mail all have an impact on the locational decisions for R&D, laboratories, and research centers.

Real Estate

Site and building costs, as well as the type, internal layout, and uses of the R&D building must be considered. Cost is generally judged by the importance of the research carried out in the buildings. More attention is given to the efficiency of the researchers and the provision of a pleasant and satisfactory environment in which the people work.

HEADQUARTERS

Communications and the *political climate* are the two most important factors in locating a headquarters for a company. Following those two factors is the availability of qualified (multilingual) personnel and the concentration of administrative headquarters. A typical example of the latter is the city of Brussels, Belgium, home to the seat of NATO, the European Commission and administration, and the European Parliament, among others.

The Political Climate

National capitals compete to attract headquarters operations. London, Paris, and Brussels are the three most attractive locations in Europe, especially for American companies. London is the main financial center of Europe and has the largest banking activities. The fact that the United Kingdom has stayed out of the European Union is a negative factor when considering London as a headquarters.

Paris was rather late in attracting American companies that wanted to set up headquarters in Europe. The rather negative attitude toward Americans has been a drawback. Further, the lack of knowledge of foreign languages in Paris has caused difficulty in establishing headquarters there. The language situation has changed in recent years, however, especially among the younger population.

Geneva, with its many international organizations might have been a competitor to Brussels. However, because Switzerland has remained out of the European Union, employs bureaucratic restrictions, and creates difficulty in obtaining permits (such as work and living permits), the ability to set up headquarters operations in or around the city of Geneva has been restricted.

Communications

Brussels is centrally located in Europe. So are Dublin, Vienna, Stockholm, and Lisbon. It follows that communications are the second most important factor in locating a headquarters.

The Brussels airport, which has grown to 20 million passengers a year, is less important than London (Heathrow), Paris (Charles de Gaulle), Amsterdam (Schiphol), or Zurich (Kloten). Although not quite an intercontinental airport, Brussels is a major European hub with a location less than an hour's flying time to the four airports mentioned.

Train connections from Brussels to the major cities of Europe are excellent. The TGV (high-speed) trains connect the center of Brussels with Paris and London, and will soon connect to Amsterdam, Cologne, and beyond.

A few decades ago, telecommunications were a national monopolistic game. They are now being privatized and modernized to follow world trends. Mobile and Internet connections are flourishing all over Europe.

CALL CENTERS

In the location of call centers, many fantastic stories can be told. If the consultant or the facility planner is Irish, he will recommend Dublin. If he is British, it will invariably be London (or even Manchester). If he's French, he will choose Paris (perhaps Lyon); if he's Dutch, Amsterdam.

The two main factors for the call center are: (1) the availability of foreign language personnel and (2) the availability of buildings meeting certain specifications in terms of size and cost.

Knowledge of Foreign Languages

English, French, German, and Dutch, in that order, are the four most important languages used in call centers. Personnel involved must be at least at a secondary education level.

Making an analysis of this factor in all major European cities is difficult since the comparison of the various institutions is purely subjective. Every nation claims to be "the best." An attempt is being made to make the requirements for obtaining degrees at European universities standard. The treaty of Bologna, for example, replaces "candidature" and "licentiate" degrees with "bachelor's" and "master's."

Saving Money on a Headquarters Study in a London Suburb

At a time when we were in the United States making a presentation of a plant location study for a large biomedical company, the president of that company learned of an allocation request for one million pounds sterling. That proposal involved building a headquarters in one of London's suburbs. The need to make that decision was urgent because the company had grown too fast and needed additional space immediately. The company offices in the United Kingdom were located on four floors of a seven-story building. Because those offices were scattered throughout the building, a rearrangement of the company's space in order to improve efficiency was urgently needed.

Since the real estate man was about to retire, he was asked to accept the challenge of dealing with the one million-pound allocation request. He wound up locating a development in which a unit costing about a million pounds could be acquired.

At this point, the company president asked us to give this project a quick look and to determine whether this one million-pound investment was necessary. We had four weeks to come up with an answer.

I flew directly from the United States to London, before returning to Brussels. In London I met three people, each frustrated about the president's hesitation for a different reason. The real estate man was hurt because the company distrusted his judgment. The manager was unhappy because the proposed project was very near his house and would reduce his current commuting time substantially. The secretary was frustrated because the current location did not offer an opportunity for noon shopping. The proposed solution was located near a shopping center.

I found no foul play in the proposal for building a one million-pound facility, but wound up reaching a very good solution for curtailing the expense. A computer company had a lease on the main floor of the company's current location that was about to expire. The two other floors not occupied by the company were leased by an insurance company that was willing to sell their space for a nominal amount.

In short, by taking advantage of the end of one lease, and buying out the other lease, the company was able to occupy all seven floors within six months. The total cost to enable the company to get the needed space all within one building was 100,000 pounds rather than one million pounds. Furthermore, the secretary decided not to quit.

The Availability of Buildings

Locating buildings that meet established size and cost specifications may sometimes be required in the siting of a call center.

* * *

This book has thus far dealt with demand. Investors need to know a lot about what is available before they can decide on a location.

The investor locates facilities in order to make money with the lowest investment of his own money and under the best conditions, thus with the least possible problems.

The following two chapters will look at the offers made by authorities (or politicians) responsible for serving the population by increasing the standard of living in an environment conducive to healthy living. *What is the role of those who make an offer to attract investments?* That is the subject of the following two chapters.

Chapter 11

Attracting Investments

REGIONAL AND ECONOMIC BALANCE

The purpose of all regional authorities, regional development offices, or foreign investment councils is *to attract foreign investment projects*. In fact, they want to attract any investment that *creates* or *maintains* jobs, be it of domestic or foreign origin.

However, the problem is that there are limited numbers of projects for many competing regions. Every region and city in Europe (and in the United States) has an agency dealing with investment attraction. These agencies all have a brochure, usually copied from one another, and nearly all making the same statements, such as:

- We are located in the heart of Europe.
- More than 300 million consumers are located within x miles.
- We have the best workers, the lowest wages, the perfect labor climate, and the best infrastructure.
- Unions? We've never heard of them.
- We offer excellent sites (some even in swampy areas) at low or no cost.
- We offer incentives, cash grants, interest rebates, tax concessions, free labor training, and the like.

Warning: the more they offer, the worse the area. Also remember: *the only thing one gets for nothing is NOTHING.* The multinationals are aware of these warning signals.

What then is the solution? Start by doing some soul searching, and understand what must be done to compete for investments in your region. Look at the following four considerations.

1. Identify the strengths and weaknesses of your own location. Honesty pays when dealing with potential clients. The companies that will consider your region or city have finely tuned skills for discovering the facts about your location.
2. Know your competitors and your competitive situation. Investors never look at only one region, unless the manager's mother-in-law lives in the area. Knowing your competitors is essential in order to make a case for your region. You should not be negative by pointing out the weaknesses of your competitors; instead, be positive about the competitive strengths of your region.
3. Identify the measures you must take in order to improve the situation in your own region; improve your competitiveness. No one is perfect. This statement also applies to regions; no region is perfect and 100 percent capable of receiving investments, but you can *strive* for perfection by getting better and better. Show a progressive attitude toward clients and point out what has been or is being done to improve certain locational factors.
4. Revise and strengthen your promotional approach. Be original in your promotions, and leave the stereotype texts to the competition.

Regional development summaries should emphasize the following four components.

1. Analysis of the strengths and weaknesses of the physical and technical elements of the location
2. Competitive analysis, dealing with the locational factors related to costs (recurring and nonrecurring)
3. Creation of instruments and policies to reach your goals, and preparation to attract investment projects suitable and desirable for your region
4. Analysis of the benefits related to macroeconomic costs by placing the region in a larger context, whether within one country or internationally

Our Experience with the Regions

The regional development offices have only one aim: to attract as many investments as possible in order to justify their existence. They are capable of using, manipulating, and restructuring facts and figures. They can present the rosiest picture possible of their region. Therefore, what they say is not so important; rather, it's what they don't disclose that is important. To illustrate our contact with the regional development offices, let us say that we never made a friend at one of those offices, nor were we considered useful to them.

Indeed, if sites were proposed in three regions, only one site in one region could and would receive investment. By favoring one region in our selection process, we made enemies of the other two regions that were not chosen. The people in the regional offices of the losing areas would go behind our backs to the investor to try to kill off the decision to select another region for its investment.

The third region, the one you recommended, will not like you because the regional development board and the local politicians will claim that it was their intervention that persuaded the investor to choose that region.

We have come across various ways in which the regional development board tried to reach its goals. They collaborated with construction companies, used law offices, and offered fake studies to consultants to get attention and favors.

We will not give specific examples although we have many in our files. However, we can say that one should beware of the smooth operators.

WHAT INDUSTRIES TO ATTRACT: MATCHING OFFER AND DEMAND

Matching supply and demand is the result of an assessment of strengths and weaknesses. *Supply* includes the locational factors offered by the regions; *demand* includes the investor's basic criteria for choosing a location.

Physical and Technological Location Factors

Matching physical and technological location factors is accomplished by considering, on one hand, the (strong) factors indicated by the regions, and on the other hand, the demand of the investors, summarized in the basic requirements of the investment project. This comparison is generally based on, but not limited to, the following six elements.

1. A general evaluation of the economic, political, and social climate
2. The investment climate in terms of social tranquility, workers' attitudes, cooperation, incentives, and taxation
3. Communications and telecommunications
4. Living conditions
5. Community facilities (educational, recreational, administrative, etc.)
6. The physical and economic elements of the sites

The matching of location factors requires analyzing the same set of criteria as listed in Chapter 8 (see Community Analysis, page 101).

The Competitive Analysis

When comparing the projection of an investment in one's own region to competing regions, one must make the same analysis that a company's management would usually make before arriving at a location decision. Calculate the following.

- Investment costs: establish cost differentials, especially in terms of magnitude, but don't try to be too precise
- The best method of financing the project
- Cost factors that affect the cost of operations: labor, freight, utilities, taxation, and incentives
- Interest payment on borrowed money
- Net profits
- Cash flows
- Investment return, payback period of one's own money, and other financial data

What Will Be the Result?

The sensitivity analysis of physical and technological elements, as well as analysis of costs, will permit the agencies to:

- Create instruments to improve their competitive position
- Establish sound regional development policies

Analysis of the Benefits Related to Macroeconomic Costs

This is a useful tool in the struggle to beat the competition. On one hand stand the *costs* of regional development policies, such as incentives provided, infrastructure, sites and site preparation, promotional material, and personnel. On the other hand stand the *benefits* of regional development policies, such as employment creation, taxation, payment balance, and subcontracting.

How to Find and Attract Investments

Assistance with promotion and acquisition of investment projects is generally defined in the development of marketing campaigns.

- Finding the needle in the haystack is not easy; one must know the industry and service sectors that are expanding.
- Check the evolution of the attitude of corporations toward your country and region.
- Don't sit and wait for the business to come to you; go after the client.
    — Find a client with a location problem. You may need to check with dozens of companies to find a client that recognizes the location problem and is setting up a project to deal with that problem.
    — Meet with the decision maker of the company. Many people in a company will tell you that they make the decisions. That makes them feel important, but don't let them fool you. Large corporations have facilities planners; smaller companies tend to pass the assignment to a vice president of finance, marketing, or sales. Sometimes even the president will keep the project close to the chest.
    — Make a proposal to decision makers showing that you understand their problem, how they make money, and what they are looking for. Be honest. If your area does not fit their needs, say so!
    — Make a clear agreement to help the client, outlining what services you will provide free of charge and those for which you will charge. Don't make promises you cannot keep.
    — Inform the client of support services provided by other agencies in your area. Don't try to provide all the services on your own. Environmental, zoning, planning, and other consider-

ations have made today's location decisions more complicated than those in the past, and a project will require other service agencies to provide that assistance.

> **Advice on How to Find Clients**
>
> In the spring of 1960, as I was working in New York City, I was invited to speak at a luncheon of the Bankers Club on Wall Street where a number of American industrialists were in attendance. Among those industrialists was the well-known chairman of a large American bank.
>
> When I asked his advice on how to find a prospective customer, he took me to the window of the room situated on the thirty-third floor, and pointed to the "ants," the many people passing the building on Wall Street. He said, "Don't just stand there—go down there, talk to as many people as you can, tell them a story about your business, and I am sure that at the end of the day you will have a client." I never had better advice.

HOW TO COOPERATE WITH INDUSTRY

Assisting in creating jobs in your area is all-important, but so is *job retention*. Cooperation with industry is not a one-time job, and industry support is particularly important at the time when a client is in the decision stage of his project.

Investment projects need continuous support and information, not only during the implementation of the project, but also during later stages of expansion. Innovative and creative ideas should be received warmly.

In my city (Mechelen, Belgium), when a large multinational company employing about 1,000 people decided to add a new product line, it was faced with the problem of wastewater treatment. Since the site was too small and long delays were encountered in obtaining permits, the authorities decided to have the treatment plant built on city-owned land, with defined specifications, yet still under the supervision of the expanding company. Upon completion of the wastewater treatment facility, the city turned the facilities over "at cost" to the company. As a result of this favorable action by the local authorities, several hundred workers were added to the local labor force.

In some companies the Chamber of Commerce will take over the role of the development agencies and keep abreast of the problems and desires of local industry groups.

THE TEN COMMANDMENTS FOR REGIONS DESIRING TO ATTRACT INVESTMENTS

1. The region's offer and the investor's demand determine how best to attract investments and combat a region's unemployment. The region must understand and analyze what it is capable of offering (sites, labor force, education, training, infrastructure, incentives, and so on).
2. The region should study competing regions offering similar advantages, especially when nonsubsidized regions must compete with those of the heavily subsidized European Union.
3. The region must know its economic, political, and financial trends.
4. The region must know the state of its regional and local economy and the connections that could result from policies to attract new investments.
5. For a projection to be efficient and to avoid wasting money, the region must know the evolution and the current situation of the economy of the areas from which new investments can be expected to make their market (e.g., Europe, United States, China, or the Far East).
6. The region must closely follow the evolution of modern technology, especially with regard to biomedical, electronics, informatics, telecommunication, automotive, robotics, media, and telematics.
7. The region must know how to prepare to receive new projects and understand the needs (i.e., the infrastructure, labor, and administrative cooperation) of those projects. The region must also understand that locations are selected and investments are made in order to make money in a peaceful social environment.
8. The region must play an important role in harmonizing human resources in terms of training, mobility, and hiring.
9. The region must assist in the harmonization of social attitudes and the relations among workers, unions, and management,

and avoid at all cost social turbulences and strikes, which will harm the region's image.
10. The region must prepare the constructive mentality of its population; the economy is in constant evolution and continuous adaptation is necessary to guarantee the population an increasing and satisfying standard of living.

Chapter 12

Industrial and Development Studies

THE EUROPEAN UNION

The European unification process is gradually affecting regional development policies. Until recently, the European Union consisted of the following fourteen countries: Belgium, the Netherlands, Luxembourg, the United Kingdom, the Irish Republic, France, Germany, Italy, Spain, Denmark, Sweden, Austria, Greece, and Finland.

On May 1, 2004, another ten countries joined. They include Cyprus, the Czech Republic, Estonia, Hungary, Latvia, Lithuania, Malta, Poland, Slovakia, and Slovenia.

The total population of the European Union is currently 387 million people, compared to 266 million in the United States and 126 million in Japan. These are the three leading economic blocs. Europe's population density is 105 people per square kilometer, compared to 29 people per square kilometer in the United States and 354 in Japan. (The *Eurostat Yearbook,* a statistical eye on Europe, provides the basic statistics on Europe and its constituent countries in comparison to the United States and Japan.)

Until recently, the European Commission was divided into twenty-one directorate-generals (DGs). Recently the administration was revised, so the new administrative composition is as follows:

1. General Services
   - Eurostat
   - Press and communications
   - Publication office
   - Secretary-general

2. Policies
    - Taxation and customs union
    - Competition
    - Economic and financial services
    - Energy and transport
    - Center for research
    - Health and consumer protection
    - Information society
    - Internal market
    - Justice and internal affairs
    - Agriculture
    - Environment
    - Enterprises
    - Education and culture
    - Research
    - Regional policies
    - Fisheries
    - Employment and social affairs
3. Foreign relations
    - Foreign relations
    - European Community Humanitarian Office (ECHO)
    - Common service for foreign affairs
    - Commerce
    - Development
    - Expansion
4. Internal Services
    - General inspection of the services
    - Budget
    - European office against fraud
    - Financial controls
    - Common translation and conference services
    - Juridical center
    - Personnel and administration
    - Translation office

Industrialists will inevitably deal with one or more of these bureaucratic sections, especially competition and regional policies. Incentives are restricted by the European Union in order to avoid falsifying information, something that seems to take place regardless of restrictions. Some regions are allowed more incentives than others. However, the regional and national governments always remain a liaison between industry and the Commission.

In Brussels, a mobile European Parliament is sometimes located in Strasbourg. This is a waste of time and money. The European Commission has thus far been independent, imposing its directives and controlled only by the ministers of the member country governments. Recently, however, the European Parliament, thus far only a figurative body, has increased its role in controlling the policies and budget of the Commission.

A new constitution is being proposed to the twenty-five member states, and must be accepted by each of the countries. This will take some time, and the outcome is uncertain.

OTHER AREAS IN EUROPE

Since the fall of the Berlin Wall more than fifteen years ago, the countries of Central and Eastern Europe have switched from a communist bureaucratic form of government to a more democratic system. The transition has been faster in some countries than in others. Hungary, the Czech Republic, Slovakia, and Poland have moved faster in aligning their economies than the other new members.

These countries are receiving more and more investments, although the transition leaves them with large unemployment problems. Bulgaria, Romania, and the Balkan countries have stayed behind the others and will require more time to adapt to the free market economic systems. Russia is still in turmoil and will probably be the last to align with Western economic policies.

THE EURO

During the Messina (Italy) meetings in 1955 when the first six countries (Belgium, Netherlands, Luxembourg—the Benelux countries—plus France, Germany, and Italy) created the European Common Market, some of the discussion focused on the creation of a common currency. Germany was still weak after its defeat, the British pound was under pressure, the French franc was being devalued constantly, governments were replaced continuously, and political instability was a fact.

It was not until more than forty years later, with the Treaty of Maastricht, that a common currency—the euro—was accepted with a fixed currency conversion. Two countries abstained (the United Kingdom and Sweden) while one country, Greece, was not accepted into the monetary union. The exchange parity was set as seen in Table 14.1.

On January 1, 2002, the national currencies disappeared and the euro became the official currency of Europe.

The advent of the euro has the ability to be a stimulant for the economies of Europe, now free of devaluations, currency exchange costs, and monetary instability. This change has also reduced some interest rates and public debts.

TABLE 14.1. Euro Equivalents

Country	Rate	Currency
Belgium	40.3399	Belgian francs
Germany	1.95583	German marks
Spain	166.386	Spanish pesetas
France	6.55957	French francs
Ireland	0.797564	Irish pounds
Italy	1936.27	Italian lira
Luxembourg	40.3399	Luxembourg francs
Netherlands	2.20371	Dutch guilders
Austria	13.7603	Austrian schillings
Portugal	200.482	Portuguese escudos
Finland	5.94573	Finnish marks

Chapter 13

Industrial Location As a Science and Its Role in the Future

INDUSTRIAL LOCATION AS A SCIENCE

For five decades we have tried, as consultants and university professors, to mold "location" into a science. The scientific method consists of evaluation, analysis, and synthesis of a given problem: Where (on what site) to locate an economic phenomenon, be it a factory, a warehouse, a research center, a call center, or a headquarters?

It was clear from the beginning (a PhD dissertation at the University of Chicago), that geography alone would not be sufficient to understand or explain a project as an economic phenomenon. As experience accumulated in dealing with management's desire to achieve maximum profit (among others) by operating on the *right site,* in the *right area,* under the *right conditions,* other sciences became involved. The first of these was economics, then factor analysis and the need for clear understanding of *demand* (by the industrialists) and the *offer* (of the public) introduced the sciences of statistics, sociology, and financial services.

Nevertheless, the previous chapters did not intend to engage in an academic discussion, but instead to show *how to solve practical everyday problems via locational tools.*

Real estate, as part of the total investment, has gradually taken a dominating role in the locational evaluation. And let us not exclude the "women" factor in the unscientific location decision, since its importance goes far beyond economic factors.

Scientific locational methodology provides the procedures, experiences, and the tools for management, not only to avoid expensive pitfalls, but also to ensure positive results. Therefore, it is understandable that the social sciences (geography, sociology, economics, statistics, and demography) provide a good background for this field.

LOCATION STUDIES AND THE FUTURE

In the 1950s and 1960s it was rather simple to do industrial site location work in Europe. Reconstruction after the devastation of World War II provided many opportunities for this.

Now, at the beginning of the twenty-first century, the selection of areas and sites has become more complicated. Conversely, more data—some questionable—are available by way of the Internet and publications of several hundred agencies.

In addition to the greenfield option, mergers, and acquisitions, other types of expansion include *clustering*—the cooperative effort of subcontractors usually squeezed out by original equipment manufacturers (OEMs).

Central and Eastern European countries that have high unemployment, low wages, and low productivity, yet a willingness to keep up with the West, have become the primary targets for European and American investments.

Opportunities in Asia, despite those elsewhere, represent the future for market share and profits. Services are gradually replacing manufacturing as the main employers in developing and developed countries.

The three dominant powers competing for market share and profits in the world are the United States, Europe, and Japan.

Monetary stability is the result of continuous economic growth in the United States and in Europe. Japan is still trying to solve its stagnation problem, although recent signs indicate improvement.

Because of globalization, corporations need to include strategic location planning as a priority. In boardrooms throughout the world, this book will rise in importance. The role of industrial location will increase in the future, and will need more attention and greater study.

We have often been asked, "If you have been working for more than a half-century in the field of 'industrial location,' have trained almost 500 students from all over the world at the University of Brussels for fifteen years, and have successfully completed, with your staff, some 1,500 job assignments in your consulting company over a period of twenty-nine years, why aren't there more people available in this field, and why are there not more consultants dealing with the subject?" My answers are:

1. First, many people call themselves "location consultants," but few have any training or experience. They think that "location" means "real estate." Most so-called location consultants mess up location decision jobs causing companies facing this problem to be reluctant to turn the assignment over to another location consultant.
2. An honest specialist in location work is *not* very popular. Why? As indicated in this book, you should not present one site or one solution to a client. In general, if a choice of four sites or communities in one or more countries is offered to a client, the result will be *three enemies* and *one friend*. The three unchosen communities or regions will *not* accept or be willing to understand that the client selected a site other than their own.
3. Do you think the fourth site, the one your client chooses, is your friend? Yes, for some time. Soon, however, the bureaucrats and politicians will start fighting for the honor that comes with getting the region selected. By the time the first stone is laid or the opening ceremony takes place, you will be long forgotten. After all, the client will have paid you your fees. This fits the wise words spoken to me in Chicago, years ago, by Milton Friedman, an economist: "I'm a man whose advice is never followed, but I still get paid for it."
4. Nevertheless, understanding the needs of a company, as well as its planning and business philosophy, is not an easy task and requires a lot of maturity. Too often consultants send inexperienced "green birds" to face weathered managers with locational problems.
5. It is difficult to remain honest if opportunity arises to make a fast buck. I have come across some strange stories. One U.S. company requested $200 thousand worth of studies from a country they were considering as a site. What the authorities ignored was the fact that consultants had assignments from two different companies to locate a plant on the island. Only one company made the decision in favor of the island. The other company withdrew because of market reasons or because it found "greener pastures" elsewhere. Or maybe, the company was tipped off.

Chapter 14

The Consultant's Role in Site Selection

SITE SELECTION PRINCIPLES

Some years ago a pharmaceutical company located a labor-intensive plant in one of the highest labor cost areas in Belgium. After four years of operation and losses, headquarters made the decision to shut down. However, fear of militant unionism, occupation of the plant, destruction of the delicate installation, and the high cost of closing, equal to the amount of the investment, all frightened the management into reconsidering. A study revealed that by changing the product line from a low-paying female-intensive labor cost operation into a high-paying male- and capital-intensive labor cost operation, both the union demands and the need for closing could be avoided. The product lines of the company were studied and several met the requirements. Within one year, the company's operation left the red and soon ran well into the black, recovering past losses.

This introductory real example shows that finding and deciding upon the right site is one of the most difficult decisions a company's management faces, be it an industrial or commercial investment. Location on the wrong site may well make the difference between profit and loss, and will definitely affect the return on investment.

Almost thirty years of experience have taught us that, if five principles are followed, the optimal site for any investment can be identified and acquired.

The first principle for making a sound investment on a site is the need to make a profit. The second principle is to achieve the highest possible return on an investment. The third principle is to bring em-

---

Paper presented at the Interindex Symposium, 1979, Basel, Switzerland, by Plant Location International (offices in Brussels, Paris, San Juan, Teheran, New York, Sydney, and Tokyo).

ployment stability to the local community and to help balance its economy.

The fourth principle is to protect the environment, both physical and human. The fifth principle is to integrate any investment in the area as soon as possible to achieve a perfect, long-lasting marriage between the industry and the community in which the site is selected.

No one can achieve long-lasting success unless *all five principles* are respected. Some have tried to respect a few and have gotten away with it for a short time only to find out that the fast buck made did not bring the happiness they expected.

Since management is not always aware of the value of these five principles, it is the duty of the consultant to defend them and to show industrial and commercial clients their value and the pitfalls if they are not followed.

WHO SEEKS SITES?

Anyone wishing to produce, distribute, or sell a product or a service, or anyone needing a research and development laboratory, headquarters, or a simple store must first find a site on which to build the facility.

Since the expansion of the American multinational companies in Europe that started in the late 1950s and peaked about the time of the 1973-1974 recession, followed by the phenomena of the increased industrial investment of European and Japanese companies in North America, one is likely to limit site seekers to only the European, American, and Japanese multinational industrial companies. Although they are the most numerous and obvious examples, site seeking is not their monopoly.

- A well-known street in Brussels is lined on both sides with cafes and small hotels. On one side, almost all beerhouses, restaurants, and small hotels are profitable; on the other side, only a few are left and are struggling to remain in business.
- Several supermarkets in the Brussels area are either having a hard time making their investments pay or are losing money outright.

- A research center in southern France in one of the most beautiful areas of the Cote d'Azur has serious problems in keeping its researchers, even at inflated wages.
- A recently established headquarters of a large multinational company is gradually losing its best personnel.

All four of these examples suffer from two deficiencies: they are located on the wrong site or the site is located in the wrong area.

WHEN A COMPANY UNDERTAKES ITS OWN LOCATION STUDY

Following are some expressions in conversations with company personnel in charge of locating new facilities.

> I will always prevent a consultant from coming in since this would put an end on my regular trips to Europe, which my wife and I enjoy so much.
>
> —American in Cleveland

> I would never trust a stranger with the information the company must reveal for a location study.
>
> —President of a European company

> We know more about our business than anyone, including our competitors, and we are not likely to tell anything to anyone.
>
> —Chicago company

> I want to control the location of the new unit, since I will be managing it and my wife is very choosy about where she wants to live.
>
> —American manager in Brussels

> Site engineering is all that counts. Of course I realize that labor and unions are important, but since our engineers will ultimately decide upon the site, we must take it to be the major element.
>
> —American engineer-planner in charge of a project

Sure, I made a mistake in Spain, but I will not make the same mistake again in Italy.

—Two-time loser of a large American company

We can go on for some time quoting examples of conversations with hundreds of prospective clients across three continents.

The examples given here *do not* suggest that companies do not have the qualified manpower to make site selection studies. It is a pleasure to discuss a locational problem with a professional. He or she understands the magnitude of the difficulties and the dangers of the pitfalls. Many companies have excellent teams in the field searching for and identifying new sites.

The examples *do* suggest, however, that not everyone is a location specialist merely because he or she was once or twice asked by the management to look for a new site.

HOW DOES THE CONSULTANT COME IN?

When a product does not sell because its *manufacturing cost* is too high, the manufacturing engineer is fired and replaced. When a product does not sell in spite of its *qualities* and *price,* the sales manager gets the axe. When *costs* cannot be properly established, the financial manager is out of a job. But who gets the blame if, four or five years later, it becomes obvious that the new plant is located on the wrong site?

The locational consultant cannot afford to make a major error in deciding the siting of a plant. Reputation, name, and company are on the line. What qualities should a location consultant possess in order to avoid making mistakes?

THE SIX MAJOR QUALITIES OF A LOCATION CONSULTANT

1. The consultant must be *objective* in evaluating the company's problem as well as the community's (or communities') qualities which match the investment profile. The consultant cannot be sidetracked by one manager's views, likes, or dislikes.

Some years ago when interviewed in London, a reporter asked me, "What do you consider the most important factor in determining where a factory will be built?" I did not hesitate to reply, "The manager's spouse." Indeed, the influence of the manager's spouse should not be underestimated. Unfortunately, by the time the plant is built, the influencing spouse and the manager might be kicked upstairs or out. "The governor of the province put us up in his chateau and during a game of golf, I promised that we would locate the new plant in his province," said the vice president of one company. The governor retired. The plant was built and closed down three years later after several million dollars were lost. There was no word from the vice president.

2. The consultant must be *thorough* in evaluating the locational factors affecting the new investment. He or she must take the critical factor analysis seriously and precisely establish the order of magnitude of both *investment* and *operational costs.* The consultant must know how the company makes money, how it operates, and what the likely environmental problems are. He or she must thoroughly study the community profiles, which are most likely to be candidates for the new investment. Finally, the consultant must analyze the preferred sites of the prospective communities or towns.

3. The consultant must be *loyal* to the client for which the study is undertaken. It pays to be honest. A reputation is acquired after more than a decade of hard work and devotion but can be lost in a few moments when greed takes over. A location consultant works *with* communities and authorities but does not sell *to* communities.

4. The consultant must *work fast.* Sometimes management takes years to make the decision to invest, but once this decision is made, they want quick answers and immediate action.

5. The consultant must keep his or her *mouth shut.* Secrecy is an all-important quality the location consultant must have in order to succeed. The client wants to stay ahead of the competition; therefore, the consultant does not want to show his or her hand. It is necessary to keep unwanted meddling between divisions to a minimum, and to shield the project from any kind of interference that would affect the locational decision through internal or external politics. The consultant wants to avoid wasting time

and needless solicitations. During the peak of the American industrial invasion in Europe, all the European Economic Countries' (EEC) ambassadors were constantly calling upon the presidents of the "big hundred," trying to influence the location of new plants. They also called upon the sales teams of countries, regions, and cities—usually needlessly.

6. The consultant must be a *diplomat,* a negotiator with strength, who knows his or her business inside out and can put forward the client's case to the authorities when applying for permits, options, incentives, or any other of the more numerous bureaucratic requirements to be satisfied.

INCENTIVES

Although it is the object of a separate paper, it is nevertheless appropriate to discuss the consultant with regard to incentives. Many times over, this book stresses that the amount of incentives offered are a factor in the location of investment. But they are not necessarily *the* determining factor. The highest incentives are offered in areas with the most problems. To lure industries into such regions will not solve the industries' or the areas' problems in the long run.

Incentives, at one time a hot topic, tend to be outweighed by operational cost, especially labor cost. However, times are changing. Europe no longer attracts labor-intensive industries that are influenced by incentives. Now, it is the capital-intensive industries that need incentives most. Profit is a dirty word, the use of which is on the decline. Not enough cash is generated quickly enough to provide for new investments. The stockholders want their share.

Environmental laws have become demanding, strict, and costly. Even cash-laden companies can no longer afford to overlook the goodies offered in many countries.

Front-end incentives are most popular. Cash grants are desired to reduce the use of one's own money or the amount borrowed. They are also used to increase investment return and to decrease the investment risk period. Fast depreciation, especially total depreciation of equipment during the year of the purchase and depreciation of half the building in the same period, is a popular incentive. Profit tax exemptions have become less attractive.

The location consultant must be able to determine the most appropriate and optimal amount of incentives the client needs to be successful on the preferred site. The area offering the largest cash grant is not necessarily the preferred area if it limits the borrowing capacity of the company and increases its required funds substantially.

More frequently, incentives will be designed specifically for certain types of industries. It is obvious that high technology electronic industries require different incentives than chemical or metalworking industries. For too long, incentives have been investments in buildings and equipment. The future calls for incentives such as brains (knowledge), productivity increases, better training, and improved skills, especially in the industrialized countries which tend to be rich in smarts and poor in raw materials. The location consultant dealing with the spectrum of specific and diversified industries is better equipped to advise on the package, both in size and type, of incentives to be proposed to the investor, without wasting the taxpayers' money.

THE FINAL RECOMMENDATIONS

The report presentation and subsequent discussion, together with visits to the preferred sites are very important parts of the consultant's task. He or she is not trying to sell a solution, strategy, or preferred site. The consultant is merely moving quickly through the rational process that was used for arriving at the conclusions.

The consultant will suggest, not impose, and will offer alternate location solutions for alternate strategies. The consultant ends up recommending a series of sites depending upon the long-term approach and strategy of the client.

IMPLEMENTING THE DECISION

The consultant plays a vital role in implementing the location decision. In general, two sites, preferably in two separate countries for negotiation leverage, will be retained.

Then, the painful and time-consuming process starts: on one hand, the multinational division must convince executive management to provide money for site acquisition and project investment. This can take several months. On the other hand, the many permits, incentives, and land acquisition must be negotiated, and this will take time. Unfortunately, it takes longer and longer as bureaucratic grips on society increase. In fact, it is an almost incomprehensible record. As the knights fight the battle, the number of permits increase and waiting time becomes longer.

During this interim period, the consultant will play a vital role in using (to the fullest) one of his or her aforementioned qualities.

IN CONCLUSION

The reader will understand that because of the high requirements in terms of knowledge, experience, and other qualities discussed, there are few acceptable locational consultants. It is hard to say if the scientist should also be an artist. A bit of imagination and much understanding and hard work are the ingredients that, when mixed together, can make a good location consultant.

Unless one can solve the clients' locational problems, and earn fees many times over, one cannot be successful and build a reputation with the large industrial companies. As in paintings, a fake is easily recognized.

The Language of Location

location: refers to the geographical position of a facility that has an economic function, such as an office, warehouse, plant, laboratory, etc.

greenfield site: land on which no project has been located before. It generally lacks infrastructure, in German referred to as *entschliessing* (developed).

brownfield site: land on which industry has been located before, such as land evacuated by abandoned steel and mining sites or chemical industry sites which—in general—are heavily polluted and carry a high cost of cleaning up, a process increasingly imposed by "green policies" everywhere in the world.

optimal location: refers to the location that meets the largest amount of criteria imposed by the clients and still obtains the highest profit under the best conditions.

logistics: formerly only used in terms of transport (cost and method); is now extended to include warehousing and handling.

benchmarking: a new term used in reference to evaluating corporate behavior with respect to location decision. It investigates corporate decisions used (and abused) in different ways to approve or disapprove a locational factor.

critical factor: refers to a factor in the locational investigation that leads to the exclusion of an area or site for further consideration.

capital intensive: refers to the total investment cost of the project divided by the number of personnel employed, resulting in a figure of more than several hundred dollars per employee. This is generally the case for chemical, steel, automotive, and other large-unit industries. Labor cost will play a smaller role in site selection.

labor intensive: refers to the total investment cost of the project divided by the number of personnel employed, resulting in a figure of less than one hundred dollars per employee. Labor cost will play a larger role in site selection. In general, this will apply to small- and medium-sized industries (SME) or large employers in the textile or electronic fields.

intangible factor: usually refers to a factor that cannot be calculated, but is likely to extensively affect cost, and therefore the locational decision. Over the years, the number of intangible factors identified has reached more than twenty.

weighted centers: geographic points of lowest transport cost. In terms of freight calculation on the basis of distances and volumes (or number of trucks) one can calculate the weighted center by measuring the distances to various shipment points and multiplying each point by the volumes (or number of trucks) shipped. The mathematical average projected on the map will mark the location point (in terms of coordinates).

Max Weber (1864-1920): a famous German historian, sociologist, and lawyer, turned economist. He taught at several universities including Berlin, Freiburg, Heidelberg, and Munich. He was the first to demonstrate the importance of intangibles such as ethical and religious factors for the development of modern capitalism.

penalty: with regard to weighted centers, indicates the higher freight cost of different geographical sites over the lowest cost determined in the weighted center.

Anjer Revolution: refers to the peaceful revolution that took place in Portugal in 1974 and ended the Salagar dictatorship period.

List of Stories

Plant Location in Spain	3
Doing It Yourself	36
The Most Important Factor in a Location Decision: A Woman	38
Ever Heard of the Mafia in Reference to Location Decision?	40
"Steal, Borrow, or Buy"	55
A Mother-in-Law Story	61
Building Construction—Story #1	63
Building Construction—Story #2	64
Water and Waste	74
Sociology and Human Relations—Story #1	84
Sociology and Human Relations—Story #2	85
Sociology and Human Relations—Story #3	86
Political Interference—Story #1	87
Political Interference—Story #2	88
Political Interference—Story #3	89
Political Interference—Story #4	89
Importance of Infrastructure—Telephone Lines	98
Matchmaking in Brussels	101
Hiding the Deficiencies of a Site	108
Locational Site Selection Bias	110
Japanese Decision Making—Story #1	110
Japanese Decision Making—Story #2	111
Promises, Promises, Promises	115
Choosing a Consultant—Story #1	118
Choosing a Consultant—Story #2	119
Choosing a Consultant—Story #3	119
Headquarters Location and Warehousing for Spare Parts	122
Saving Money on a Headquarters Study in a London Suburb	126
Our Experience with the Regions	131
Advice on How to Find Clients	134

Location, Location, Location
© 2008 by The Haworth Press, Taylor & Francis Group. All rights reserved.
doi:10.1300/6094_17

List of Location Tools

Project Requirements Questionnaire	26
The Ten Main Intangibles	41
The BERI Rating	42
Incentives: Cash Grant	65
The Investor Climate	81
Community Analysis	101
Site Analysis	104
The Ten Commandments for Regions Desiring to Attract Investments	135
The Consultant's Role in Site Selection	145

Index

Page numbers followed by the letter "t" indicate tables.

Acquisitions, 53-54
Area concentration, 11-12
Area of search, 8, 46-47, 51
Arthur Andersen, 22
Asia, 142

Basic requirements, of location project, 7
Benefits analysis, related to macroeconomic costs, 133
BERI rating. *See* Business Environment Risk Intelligence (BERI) rating
BLS. *See* Business Location Services (BLS)
Bristol-Myers, 45-46
Buck, Rene, 23
Buck Consultants International, 23
Building construction example, 63-64
Building costs, 61-64
Business Environment Risk Intelligence (BERI) rating, 41, 42, 42t
Business Location Services (BLS), 22

Call centers, locating
  building availability, 127
  foreign language knowledge, 125-126
Capital intensive project, 39

Cash flow tables, 93-94, 94t, 95t
Cash grants, 65-67, 79t, 150-151
Clustering, 117, 142
Community analysis, 100
  community facilities, 103-104
  employment, 103
  industrial site, 104
  industrialization, 102-103
  location, 101
  population, 102
  transportation, 101-102
Competitive analysis, 132
Competitive position, 70
Compulsory purchase powers, 49, 58
Computer programs, 50
Consultants, 22-23
  avoiding mistakes, 148
  example, 118-120
  qualities, 148-150
  site selection role
    companies undertaking location studies, 147-148
    conclusion, 152
    implementing decision, 151-152
    incentives, 150-151
    principles, 145-146
    recommendations, 151
    seekers of sites, 146-147
  use of, 115-118
Conway, McKinley, 5
Cost estimate, 62

Costs
  building, 61-64
  development, 58, 59
  equipment, 64-65
  estimate, 62
  inbound freight, 50, 52, 70-71
  interplant freight, 71
  investment, 8, 17
    building costs, 61-64
    equipment costs, 64-65
    incentives, 65-67
    site costs, 58-61
  labor, 63, 69-70
  land, 58
  leasing, 59
  locational, 50-52
  macroeconomic, 133
  nonrecurring, 8, 51
    calculating investment costs, 58-67
    example, 55-57
  operational, 8, 17, 51-52
  outbound freight, 50, 52, 71
  power, 72-73
  product, 70
  recurring, 8, 51, 57
    example, 74
    labor costs, 69-70
    tax incentives, 77-79
    taxation, 74-76
    transportation costs, 70-72
    utility costs, 72-74
  site, 58-61
  transportation, 44-45, 47, 50, 121-122
  utility
    gas, 73
    pollution treatment, 73, 74
    power, 72-73
    water, 73, 74
Critical factors, 7, 15
  analysis
    acquisitions, 53-54
    area of search, 46-47
    example, 47
    factor types, 43-46

Critical factors, analysis *(continued)*
  feasibility studies, 53
  joint ventures, 53-54
  local recommendations, 48
  locational costs, 50-52
  mergers, 53-54
  physical fit, 49-50
  weighted centers, 50, 52-53
  detecting, 34-35

Decision making
  conclusions, 111-114
  consultants, 115-118
  example, 110-113, 118-120
  recommendations, 114-115
  weighing of factors, 109-111
Demand, 3-4
  benefits analysis, 133
  competitive analysis, 132
  finding and attracting investments, 133-134
  physical and technological location factors, 131-132
  results, 132
Determining factors, 8, 35-38, 69
Development costs, 58, 59
Development studies
  Central and Eastern Europe, 139
  euro, 139-140
  European Union, 137-139
DFI. *See* Direct Foreign Investment (DFI)
DGs. *See* Directorate-generals (DGs)
Differentials
  in European Union, 77, 78t
  magnitude of, 55, 93
Direct Foreign Investment (DFI), 92
Directorate General for Employment, Industrial Relations, and Social Affairs, 83
Directorate-generals (DGs), 137
"Doing it yourself" example, 36
Dupont du Nemours, Inc., 10, 11

ECHO. *See* European Community Humanitarian Office (ECHO)
Ecology, 19, 94-95
Economic analysis
   nonrecurring cost analysis, 55-67
      calculating investment costs, 58-67
      example, 55-57, 61, 63-64
   recurring cost analysis
      example, 74
      labor costs, 69-70
      tax incentives, 77-79
      taxation, 74-76
      transportation costs, 70-72
      utility costs, 72-74
Effective tax rates, 75
Électricité de France (EDF), 73
Employment and Labor Market in Europe (publication), 83
EMU. *See* European Monetary Unit (EMU)
Entschliesungskosten (development costs), 58, 59
Equipment costs, 64-65
Ernst & Young, 22
Euro (European common currency), 24-25, 86, 87, 91, 139-140, 140t
Europe. *See also* European Union
   Central, 24, 139
   Eastern, 139
   industrial location in, 9-11
European Commission, 54, 72, 83, 86, 137-139
European common currency. *See* Euro (European common currency)
European Community Humanitarian Office (ECHO), 138
European Monetary Unit (EMU), 91
European Parliament, 24-25, 86, 139
European Trade Union Yearbook: 1995 (edited by Sabaglio and Hoffman), 83-84
European Union, 24, 72. *See also* Europe
   development studies, 137-139

European Union *(continued)*
   industrial studies, 137-139
   political stability, 84-86
   tax incentives, 77-78, 78t, 79t
   tax rates, 75-76, 76t
Examples
   advice on how to find clients, 134
   building construction, 63-64
   choosing a consultant, 118-120
   critical factors analysis, 47
   "doing it yourself," 36
   experience with regions, 131
   headquarters location and warehousing for spare parts, 122-123
   hiding deficiencies of site, 108
   infrastructure importance, 98
   Japanese decision making, 110-113
   locational site selection bias, 110
   Mafia influence on location decision, 40-41
   matchmaking in Brussels, 101
   Mechelen, Belgium location, 9-11
   mother-in-law story, 61
   plant location in Spain, 3-4
   political interference, 87-89
   "promises, promises, promises," 115
   saving money on headquarters study, 126
   sociology and human relations, 84, 85, 86-87
   "steal, borrow, or buy," 55-57
   water and waste, 74
   women's roles in location decision, 3-4, 38
Experience with regions example, 131

Facility planner, 5
Factors, in location decisions, 22
   critical, 7, 15
      analysis, 43-54
      detecting, 34-35
      determining, 7, 35-38, 69

Factors, in location decisions, *(continued)*
  intangible, 51, 52
    economic situation, 87-89
    example, 84, 85, 86-89
    investment climate, 81-82
    political situation, 84-87
    social climate, 82-84
  noncalculable, 51
  nonrecurring costs, 8, 51
    calculating investment costs, 58-67
    example, 55-57
  recurring costs, 8, 51
    example, 74
    labor costs, 69-70
    tax incentives, 77-79
    taxation, 74-76
    transportation costs, 70-72
    utility costs, 72-74
  weighing, 109-111
Figures, 38-39, 55, 64-65
Financing, of investments, 91-93
Ford Motor Company, 11, 12, 14, 35-36
Friedman, Milton, 143

Gas suppliers, 73
General Motors, 35, 46
Genk Taunus plant, 11, 12
Germany, unification, 23-24
Global Location and Expansion Services (GLES), 22
Grants, cash, 65-67, 79t, 150-151
Green political parties, 73, 74
Gross disbursement, 69

Headquarters, locating, 124-125
  communications, 125
  example, 122-123, 126
  political climate, 124
Hiding deficiencies of site example, 108
Hoffman, Reiner, 83

IDRC. *See* International Development Research Council (IDRC)
ILAS. *See* International Location Advisory Services (ILAS)
Inbound freight cost, 50, 52, 70-71
Incentives, 65-67, 79t
  front-end, 150-151
  restriction, 138-139
Industrial location
  community approach to, 8
  and ecology, 94-95
  in Europe, 9-11
  location decisions, 9
    call centers, 125-127
    consultants, 115-118
    critical factors, 7, 15, 34-38, 43-54
    determining factors, 7, 35-38, 69
    drawing conclusions, 111-114
    example, 3-4, 110-113, 115, 118-120
    headquarters, 124-125, 126
    importance, 1
    intangible factors, 51, 52, 81-89
    noncalculable factors, 51
    nonrecurring cost factors, 8, 51, 55-67
    offers and demands, 3-4, 131-134
    people involved, 1-2, 4-5
    potential mistakes, 2
    preparation, 5-6
    recommendations, 114-115
    recurring cost factors, 8, 51, 69-79
    research and development (R & D) centers, 122-124
    training, 5
    warehouses, 121-123
    weighing factors, 109-111
    women's role, 3-4, 38, 141
  location study
    area concentration, 11-12
    company undertaking own, 147-148
    example, 36
    and future, 142-143

Industrial location, location study *(continued)*
    methodology changes, 21-25
    purpose, 7-11
    questionnaire, 33-34
    supplier concentration, 12
    techniques, 12-21
  model, 11
  as science, 141-142
  site selection
    companies undertaking location studies, 147-148
    implementing decision, 151-152
    incentives, 150-151
    principles, 145-146
    qualities of location consultant, 148-150
    recommendations, 151
    seekers of sites, 146-147
Industrial studies, 137-138
  Central and Eastern Europe, 139
  euro, 139-140
  European Union, 139
Infrastructure
  example, 98
  transportation, 106
  utilities, 106-107
Intangible factors, 51, 52
  BERI rating, 41, 42, 42t
  economic situation, 87-89
  example, 84, 85, 86-89
  investment climate, 81-82
  looking at, 39-41
  political situation, 84-87
  social climate, 82-84
  ten main, 41-42
International Development Research Council (IDRC), 5, 116
International Investment Group of Price Waterhouse, 22
International Location Advisory Services (ILAS), 22
Interplant freight costs, 71
Interviews, 20-21
Investment climate, 8, 81-82

Investment costs, 8
  calculating
    building costs, 61-64
    equipment costs, 64-65
    incentives, 65-67
    site costs, 58-61
  financial assumptions, 17
Investments
  attracting
    cooperating with industry, 134-135
    example, 131, 134
    matching offer and demand, 131-134
    regional and economic balance, 129-131
    ten commandments, 135-136
  financing, 91-93
  nature of, 39
  real estate, 122

Japanese decision making example, 110-113
Job retention, 134
Joint ventures, 53-54
Just-in-time delivery, 71

KPMG, 22

Labor
  costs, 63, 69-70
  pirating, 15
  unions, 99
Labor-intensive investment projects, 39, 69
Land costs, 58
Leasing costs, 59
Location decisions, 9
  call centers
    building availability, 127
    foreign language knowledge, 125-126

Location decisions *(continued)*
  consultants, 115-118
  drawing conclusions, 111-114
  example, 110-113, 115, 118-120
  factors
    critical, 7, 15, 34-38, 43-54
    determining, 7, 35-38, 69
    intangible, 51, 52, 81-89
    noncalculable, 51
    nonrecurring costs, 8, 51, 55-67
    recurring costs, 8, 51, 69-79
    weighing, 109-111
  headquarters
    communications, 125
    example, 122-123, 126
    political climate, 124
  importance, 1
  offers and demands, 3-4
  people involved, 1-2, 4-5
  potential mistakes, 2
  preparation, 5-6
  recommendations, 114-115
  research and development (R & D)
    centers, 122-124
    people, 123
    real estate, 124
  training, 5
  warehouses
    example, 122-123
    real estate investment, 122
    transportation costs, 121-122
  women's role, 3-4, 38, 141
Location study
  area concentration, 11-12
  company undertaking own, 147-148
  example, 36
  and future of industrial location, 142-143
  methodology changes, 21
    confusion, 23
    consultants, 22
    factors, 22
    freight, 22
    historical events, 23-25
    personnel, 22

Location study *(continued)*
  purpose, 7-11
  questionnaire, 33-34
  supplier concentration, 12
  techniques, 12-21
    company policies, 13-14
    requirements, 14-19
    tools, 19-21
Locational feasibility study, 53
Locational site selection bias example, 110
Logistics, 72

Macroeconomic costs, benefits analysis related to, 133
Mafia, influence on location decision, 40-41, 60
Magnitude of differentials, 55
Marketing, 133-134
Matchmaking in Brussels example, 101
Mechelen, Belgium location example, 9-11
Mergers, 53-54
Monroe Auto Equipment Company, 12, 44
Mother-in-law story example, 61

Noncalculable factors, 51
Nonrecurring cost factors, 8, 51
  calculating investment costs, 58-67
  example, 55-57

OEMs. *See* Original equipment manufacturers (OEMs)
OM (own money), 91-92
Operational cost, 8, 17, 51-52
OPM (other people's money), 91-92
Order of magnitude, 64-65
Original equipment manufacturers (OEMs), 117, 142
Outbound freight cost, 50, 52, 71

Permits
  building, 60
  as critical factor, 45
  operating, 60
Physical fit, 8, 49-50
Plant Location International (PLI), 12, 19, 22, 23, 116-117
PLI. *See* Plant Location International (PLI)
Political interference example, 87-89
Pollution treatment, 73, 74
Power
  costs, 72-74
  supply, as critical factor, 45, 46
Price Waterhouse, 21-23
Procter & Gamble, 10
Product
  costs, 70
  as critical factor, 43-44
Project definition, 7
  critical factors, 34-35
  determining factors, 35-38
  intangible factors, 39-42
    BERI rating, 41, 42, 42t
    looking at, 39-41
    ten main, 41-42
  nature of investment, 39
  use of figures, 38-39
Project requirements questionnaire, 14-19, 26-32, 33-34
"Promises, promises, promises" example, 115

R & D centers. *See* Research and development (R & D) centers
Real estate
  research and development (R & D) centers, 124
  warehouses, 122
Recurring cost factors, 8, 51, 57
  example, 74
  labor costs, 69-70
  tax incentives, 77-79
  taxation, 74-76

Recurring cost factors *(continued)*
  transportation costs, 70-72
  utility costs, 72-74
Regional development summary, 130
Research and development (R & D) centers
  locating, 122-124
  people, 123
  real estate, 124
Russia, 24

Sabaglio, Emilio, 83
Salazar, Antonio, 85
Silicon Valley, 37
Site analysis, 97
  community analysis, 100, 101-104
  example, 98, 101, 108
  geographic aspects, 97
  infrastructure, 106-107
  site description, 104-105
  site location process, 100-101
  socioeconomic aspects, 98-100
Site costs, 58-61
Site selection, consultant's role
  avoiding mistakes, 148
  companies undertaking location studies, 147-148
  conclusion, 152
  implementing decision, 151-152
  incentives, 150-151
  principles, 145-146
  qualities of location consultant, 148-150
  recommendations, 151
  seekers of sites, 146-147
Social climate, 82-84
Sociology and human relations
  example, 84, 85, 86-87
Spain, location decision example, 3-4, 44-45
Statistics, 21, 55, 65
Statutory tax rates, 75
"Steal, borrow, or buy" example, 55-57
Subsidies, 93
Supplier concentration, 12

Supply, 3-4
   benefits analysis, 133
   competitive analysis, 132
   finding and attracting investments, 133-134
   physical and technological location factors, 131-132
   results, 132

Tax incentives, 77-79
Tax rates
   effective, 75
   statutory, 75
Taxation, 74-76
Ten commandments, to attract investments, 135-136
Trade unions, 83-84
Transportation costs, 44-45, 47, 50, 70-72
   inbound freight costs, 70-71
   interplant freight costs, 71
   logistics, 72
   outbound freight costs, 71
   warehouse location, 121-122
Trends analysis, 57

Unemployment, 99
Unions
   labor, 99
   trade, 83-84
Utility costs
   example, 74
   gas, 73
   pollution treatment, 73, 74
   power, 72-73
   water, 73, 74

Warehouses, locating
   example, 122-123
   real estate investment, 122
   transportation costs, 121-122
Water and waste example, 74
Water costs, 73, 74
Water supply, as critical factor, 45-46
Weber, Max, 50
Weighted center, 50, 52-53
Women, role in location decision, 3-4, 38, 141

Made in the USA
Coppell, TX
15 August 2020